T0189651

# Advances in Intelligent Systems and Computing

Volume 390

**Series editor**

Janusz Kacprzyk, Polish Academy of Sciences, Warsaw, Poland
e-mail: kacprzyk@ibspan.waw.pl

## *About this Series*

The series "Advances in Intelligent Systems and Computing" contains publications on theory, applications, and design methods of Intelligent Systems and Intelligent Computing. Virtually all disciplines such as engineering, natural sciences, computer and information science, ICT, economics, business, e-commerce, environment, healthcare, life science are covered. The list of topics spans all the areas of modern intelligent systems and computing.

The publications within "Advances in Intelligent Systems and Computing" are primarily textbooks and proceedings of important conferences, symposia and congresses. They cover significant recent developments in the field, both of a foundational and applicable character. An important characteristic feature of the series is the short publication time and world-wide distribution. This permits a rapid and broad dissemination of research results.



More information about this series at http://www.springer.com/series/11156

Richa Singh · Mayank Vatsa
Angshul Majumdar · Ajay Kumar
Editors

# Machine Intelligence and Signal Processing

*Editors*
Richa Singh
Indraprastha Institute of Information Technology
New Delhi
India

Angshul Majumdar
Indraprastha Institute of Information Technology
New Delhi
India

Mayank Vatsa
Indraprastha Institute of Information Technology
New Delhi
India

Ajay Kumar
Department of Computing
Hong Kong Polytechnic University
Hong Kong
China

ISSN 2194-5357 ISSN 2194-5365 (electronic)
Advances in Intelligent Systems and Computing
ISBN 978-81-322-2624-6 ISBN 978-81-322-2625-3 (eBook)
DOI 10.1007/978-81-322-2625-3

Library of Congress Control Number: 2015947800

Springer New Delhi Heidelberg New York Dordrecht London


Printed on acid-free paper

Springer (India) Pvt. Ltd. is part of Springer Science+Business Media (www.springer.com)

# Preface

## Overview and Goals

Machine learning and signal processing have been prevalent in every other research challenge. Today, various acquisition devices collect and generate massive amounts of data. Processing and analyzing such data is usually beyond our mental capacities. We resort to automated processing and analysis techniques to retrieve information from this data deluge. Several researchers across the globe from academics, industry and government agencies are working in signal processing and machine learning related problems. It is therefore important to understand the state-of-the-art approaches in related research areas. The primary goal of this book is to present some recent advances in these areas.

## Organization and Features

The book includes four invited chapters from Dr. Ajay Kumar, Dr. Vidit Jain, Dr. Amit Kale, and Dr. Girish Chandra along with 10 chapters on different problems of machine intelligence and signal processing from researchers across the world. The four invited chapters are from the speakers of the First Winter School on Machine Intelligence and Signal Processing and the 10 chapters are peer-reviewed and selected papers of the workshop. The chapters cover theoretical aspects such as the chapter on Greedy Algorithms for Non-linear Sparse Recovery by Kavya Gupta and Angshul Majumdar. It also includes application studies in different application areas such as the chapters on retinal blood vessel classification by Rajan, Ballerini, Truccco and MacGillivray, and SMS classification by Goswami, Singh and Vatsa. The authors who have contributed in this book are from academia and corporates of different countries including Hong Kong, India, Italy, and the United Kingdom.

## Target Audiences

The target audience of this book are individuals who are interested in advancements in different areas of machine intelligence and signal processing. Researchers in both industry and academia should find it useful for its breadth of topics and applications, such as medical image analysis, biometrics, and text classification.


## Acknowledgments

We would like to thank the editors at Springer for their help, patience, and advice in the course of developing this book. We also want to thank all of the contributors of this book. We want to thank all of our collaborators at our respective institutions for the vibrant research atmosphere that they have provided. Finally, we also thankfully acknowledge the support and efforts from our students in organizing the winter school, from where the chapters of this book have been invited and collected.



New Delhi, India — Richa Singh
New Delhi, India — Mayank Vatsa
New Delhi, India — Angshul Majumdar
Hong Kong, China — Ajay Kumar

# Contents

# About the Editors

**Richa Singh** received her M.S. and Ph.D. degrees in computer science in 2005 and 2008, respectively from the West Virginia University, Morgantown, USA. She is currently an Associate Professor and recipient of the Kusum and Mohandas Pai Faculty Research Fellowship at the Indraprastha Institute of Information Technology (IIIT) Delhi, India. Her research has been funded by the UIDAI and DEITY, Government of India. She is a recipient of the FAST award by DST, India. Her areas of interest are biometrics, pattern recognition, and machine learning. She has more than 150 publications in refereed journals, book chapters, and conferences. She is an editorial board member of the Journal of Information Fusion and EURASIP Journal on Image and Video Processing. She is the PC Co-Chair of International Conference on Biometrics: Theory, Applications and Systems, 2016. Dr. Singh is a member of the IEEE, Computer Society and the Association for Computing Machinery. She is a recipient of several best paper and best poster awards in international conferences.

**Mayank Vatsa** received the M.S. and Ph.D. degrees in computer science in 2005 and 2008, respectively from the West Virginia University, Morgantown, USA. He is currently an Associate Professor and AR Krishnaswamy Faculty Research Fellow at the Indraprastha Institute of Information Technology (IIIT) Delhi, India. He has more than 150 publications in refereed journals, book chapters, and conferences. His research has been funded by the UIDAI and DEITY, Government of India. He is a recipient of the FAST award by DST, India. His areas of interest are biometrics, image processing, computer vision, and information fusion. Dr. Vatsa is a member of the IEEE, Computer Society and the Association for Computing Machinery. He is the recipient of several best paper and best poster awards in international conferences. He is also an associate editor of IEEE Access, area editor of Information Fusion, and IEEE Biometric Compendium, and served as the PC Co-Chair of ICB 2013 and IJCB 2014. He is the Vice President of IEEE Biometrics Council—Publications.

**Angshul Majumdar** is currently an assistant professor at Indraprastha Institute of Information Technology Delhi. He is working at this institute since October 2012. He did his Master's and Ph.D. in electrical engineering from the University of British Columbia in 2009 and 2012 respectively. His Ph.D. thesis was on accelerating Magnetic Resonance Imaging scans using sparse recovery techniques. His current research interests are algorithms for sparse signal recovery a.k.a Compressed Sensing and low-rank matrix recovery. He applies these techniques for problems arising in biomedical signal and image acquisition. Angshul has written more than 60 papers in reputed journals on these topics. Currently Angshul is the president of the IEEE Signal Processing Society chapter under the IEEE Delhi section.

**Ajay Kumar** received the Ph.D. degree from the University of Hong Kong, Hong Kong in 2001. He was an assistant professor with the Department of Electrical Engineering, IIT Delhi, Delhi, India from 2005 to 2007. He is currently working as Associate Professor in the Department of Computing, The Hong Kong Polytechnic University. His current research interests are biometrics with an emphasis on hand biometrics, vascular biometrics, iris, and multimodal biometrics. He holds three US patents, and has authored on biometrics and computer vision-based industrial inspection. He is currently an Area Editor for Pattern Recognition Journal and served on the IEEE Biometrics Council as the Vice President (Publications). He was on the Editorial Board of the IEEE Transactions on Information Forensics and Security from 2010 to 2013, and served on the program committees of several international conferences and workshops in the field of his research interest. He was the program chair of the Third International Conference on Ethics and Policy of Biometrics and International Data Sharing in 2010, the program co-chair of the International Joint Conference held in Washington, DC, USA, in 2011, the International Conference on Biometrics held in Madrid, in 2013, CVPR 2013 Biometrics Workshop held in Portland, CVPR 2014 Biometrics Workshop held in Columbus, CVPR 2015 Workshop held in Boston, and has also served as the General Co-Chair of the Second International Joint Conference on Biometrics held in Tampa, in 2014 and First International Conference Identity, Security and Behavior Analysis held in Hong Kong in 2015.

# Advancing Cross-Spectral Iris Recognition Research Using Bi-Spectral Imaging

**N. Pattabhi Ramaiah and Ajay Kumar**

**Abstract** Iris images are increasingly employed in national ID programs and large-scale iris databases have been developed. Conventional iris images for the biometrics databases are acquired from close distances under near infrared wavelengths. However the surveillance data is often acquired under visible wavelengths. Therefore, applications like watch-list identification and surveillance at-a-distance require accurate iris matching capability for images acquired under different wavelengths. In this context, simultaneously acquired visible iris images should be matched with the iris images acquired under near infrared illumination to ascertain cross-spectral iris recognition accuracy. This paper describes the need for such cross-spectral iris recognition capability and proposes the development of bi-spectral iris image database to advance the much needed research in this area.

## 1 Introduction

Various government sectors in the world provide public services and welfare schemes for the benefit of people in the society. Iris recognition is increasingly employed for large-scale applications, especially in the government supported projects globally. Recent survey on the biometrics market [1] indicates that funding for government biometric-based projects such as National ID Program, Biometric Drivers license, Biometric Passports & Visas, is expected to be about 21.9 billion US$ by 2020. Almost all of the currently deployed iris recognition systems [2, 3] use the data acquired at near infrared (NIR) wavelengths. These systems are believed to be more accurate among all the existing biometric recognition systems. There have been recent efforts to develop visible wavelength-based iris recognition techniques in order to

N.P. Ramaiah (✉) · A. Kumar
Department of Computing, The Hong Kong Polytechnic University, Hung Hom, Hong Kong, China
e-mail: n.p.in@ieee.org

A. Kumar
e-mail: csajaykr@comp.polyu.edu.hk



R. Singh et al. (eds.), *Machine Intelligence and Signal Processing*, Advances in Intelligent Systems and Computing 390,
DOI 10.1007/978-81-322-2625-3_1

eliminate the NIR radiation hazards and the limitations of existing iris recognition systems. In this context, iris pigmentation [4] plays an important role while acquiring images under visible wavelengths. The concentration of melanin pigment is more in dark colored irises which can be acquired with greater texture details under near infrared illumination. The color of the iris can be determined by considering the variable proportions of the two molecules, namely enumelanin (brown/black) and pheomelanin (red/yellow). The appearance of dark colored iris is not very clear under visible illumination since it has more enumelanin molecules which absorb the visible illumination deeply. In order to ensure enhanced texture details under visible wavelengths, the power of visible illumination should be increased which can be uncomfortable to the users. Also, the external environmental illumination can add more noise to the visible iris images in terms of shadows, specular and diffuse reflections. These factors should be considered before selecting an image acquisition setup especially under visible wavelengths.

There have been several promising attempts to acquire the iris images under visible wavelengths. Proenca et al. [6, 7] released UBIRIS database in two different versions in which the iris data was collected in visible wavelengths by considering noise artifacts and unconstrained acquisition conditions (at-a-distance and on-the-move). Dobes et al. [8] released UPOL iris database which is captured using TOPCON TRC50IA optical device in fully constrained acquisition conditions under visible wavelengths. Tajbakhsh et al. [9] explored a new approach for iris recognition by considering the fusion of iris extracted features captured under both near infrared and visible wavelengths, but not simultaneously. Reference [10] describes a framework for accurately segmenting the iris images from face images acquired at-a-distance under near infrared or visible illuminations.

Very few efforts have been made to analyze multi-spectral iris imaging. Ross et al. [11] proposed the enhanced iris recognition system by considering the fusion of multiple spectrum beyond wavelengths of 900 nm. Recently, cross-spectral periocular matching has been explored in [12] using trained neural network with interesting results. An approach for cross-spectral iris matching [13] was proposed using the predictive NIR image. Even though the cross-spectral verification accuracy [13] was very close to the accuracy of NIR cross-comparisons, it is far from being practical because prediction is different from reality. An overview of various related iris recognition methods for different cross-comparisons is explained in Table 1. To the best of our knowledge, there is no publicly available database to address the exact problem of cross-spectral iris recognition. The solution to cross-spectral iris recognition is required to authenticate iris images acquired under visible wavelengths with large-scale iris databases in which the data is acquired in near infrared wavelengths.

The rest of the paper is organized as follows: In Sect. 2, the proposed image acquisition setup is presented in order to acquire the cross-spectral iris images. Experimental results for cross-spectral iris recognition are discussed in Sect. 3. Key conclusions and need for future research efforts are discussed in Sect. 4.

**Table 1** Summary of related work on different cross-domain iris matching

| Reference | Method | Iris comparisons | Database |
|---|---|---|---|
| [11] | Score level fusion is used for iris images acquired beyond wavelengths of 900 nm to improve the matching performance | Multi-spectral | WVU Multi-spectral iris database |
| [13] | A predictive NIR iris image is used from the color image | Cross-spectral | WVU Multi-spectral iris database |
| [9] | Comparisons are made based on the feature fusion of both near infrared and visible iris images | Cross-spectral | UTIRIS iris database |
| [14] | A sensor adaptation algorithm is proposed for cross-sensor iris recognition using kernel transformation learning | Cross-sensor | BTAS 2012 Cross-sensor Iris Competition database (ND Dataset) |
| [15] | An optimization model of coupled feature selection is proposed for comparing cross-sensor iris data | Cross-sensor | BTAS 2012 Cross-sensor Iris Competition database (ND Dataset) |

## 2 Acquiring Bi-Spectral Images for Cross-Spectral Iris Recognition

In order to meet the requirement, a GigE compliant multi-spectral camera [18] is selected which utilizes dichroic prism (see Fig. 1), one for visible imaging and the other for near infrared imaging. The precise separation of the visible and near infrared regions of the spectrum is illustrated in Fig. 2. In order to adjust both the focal points optimally, it is advisable to use any lens specially designed for 3CCD cameras. The image acquisition setup is illustrated in Fig. 3. Communication between the camera and the computer is through dual NIC, one NIC is for visible and the other is for near infrared channel. The captured visible and near infrared images have dimensions of image size, $640 \times 480$ pixels. Choosing illumination sources for both the spectrums is more significant while capturing the images since there is always a trade-off between the quality of visible and near infrared images. In order to acquire good quality iris images, factors like exposure, gain, and illumination intensity are independently considered for both the channels.

The camera specifies that the exposure limits ($\delta_e$) range from 0 to 792 and the gain limits ($\delta_g$) range from $-89$ to 593. After analyzing different exposure and gain combinations, the settings for both the channels are finalized empirically as shown in Table 2. In visible channel, the exposure and gain settings are fixed as 150 and 75, respectively. The illumination source used for the visible channel is a professional video light (VL-S04) whose illumination intensity is 1450 lx and the illumination angle is 45 degrees. In a similar way, the exposure and gain parameters were fixed as 15 and $-80$, respectively, for the near infrared channel. The near

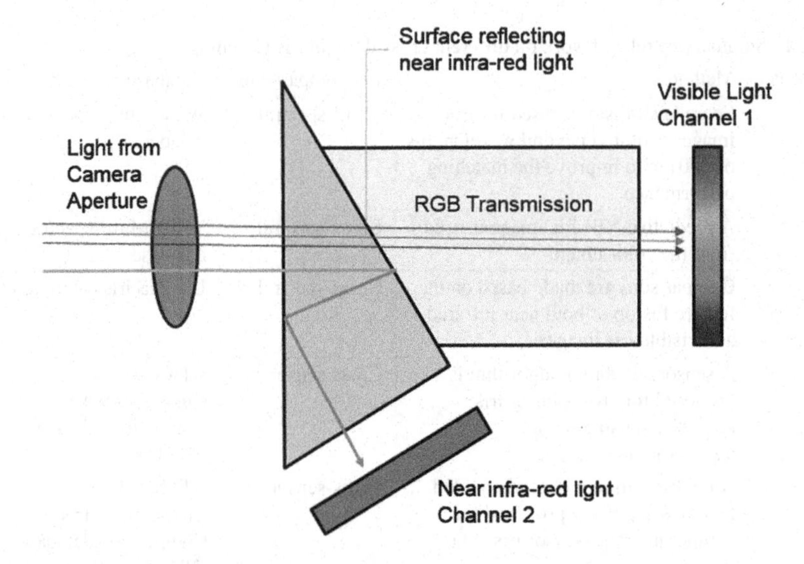


**Fig. 1** Simultaneous iris imaging using dichroic prism optics

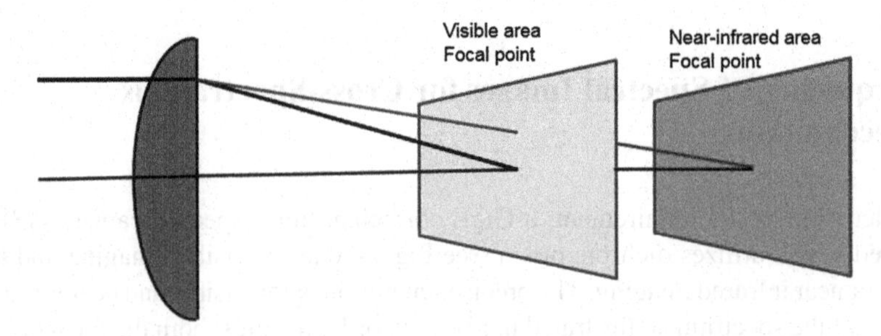


**Fig. 2** Focal points for visible and near infrared illuminations

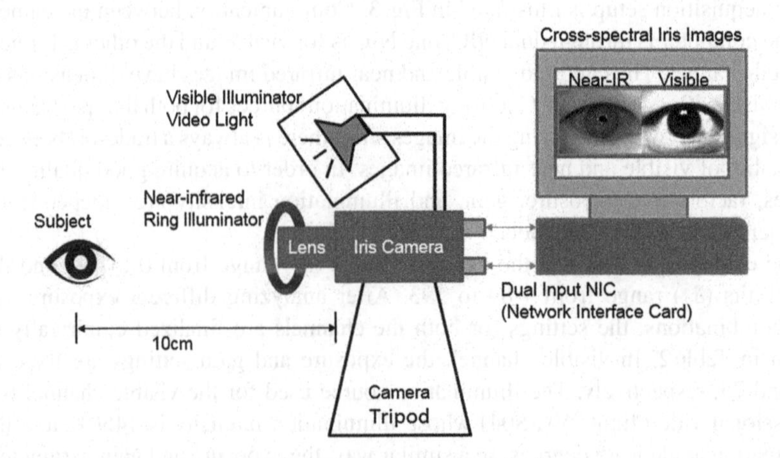


**Fig. 3** Image acquisition setup for bi-spectral iris imaging

**Table 2** Imaging parameters for the visible and near infrared channels

| Variable | Near infrared channel | Visible channel |
|---|---|---|
| Exposure limit value ($\delta_e$) | 15 | 150 |
| Gain limit value ($\delta_g$) | −80 | 75 |
| Illumination intensity ($E_{IR}$) ($W/m^2$) | 2.6 | 2.123 |
| Illumination angle (*degrees*) | 60 | 45 |

infrared illumination can cause damage to the eye if its intensity exceeds some limit. In order to alleviate near infrared radiation hazard, a ring illuminator is specifically designed which is suitable for the proposed acquisition environment. The total IR illumination intensity of 2.6 W is used at a distance of 10 cm from ring illuminator to the eye. The maximum acceptable IR radiation [19], $E_{IR}$, can be defined in the wavelength range 780–3000 nm as

$$E_{IR} = \sum_{\lambda=780}^{\lambda=3000} E_{\lambda} \bullet \Delta\lambda \leq 18000 \bullet t^{-0.75} \, [\mathrm{W} \bullet \mathrm{m}^{-2}], \, for\, t \leq 1000 \text{ s},$$
$$= \sum_{\lambda=780}^{\lambda=3000} E_{\lambda} \bullet \Delta\lambda \leq 100 \, [\mathrm{W} \bullet \mathrm{m}^{-2}], \, for\, t > 1000 \text{ s}, \quad (1)$$

where $E_{\lambda}$ is the spectral irradiance, $\Delta\lambda$ is the spectral bandwidth source in nm and t is the exposure time. The setup captures the iris images in both channels simultaneously and saves the images as well the sequence of different images at an interval of 0.5 s for further investigation. The sample iris images acquired using the proposed acquisition setup are illustrated in Fig. 4.

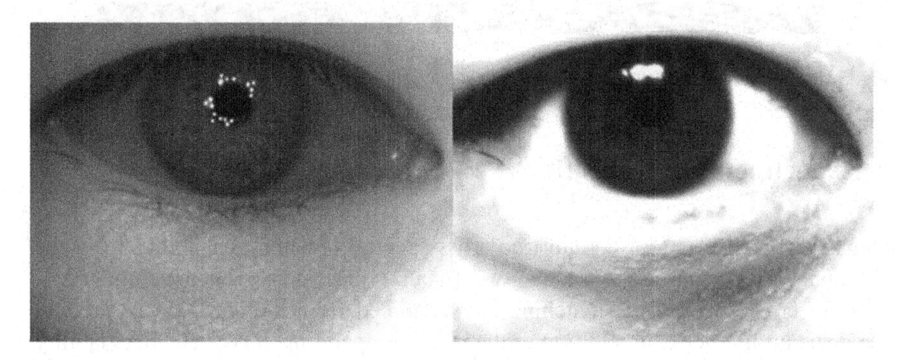

**Fig. 4** Sample iris images simultaneously acquired under near infrared and visible wavelengths

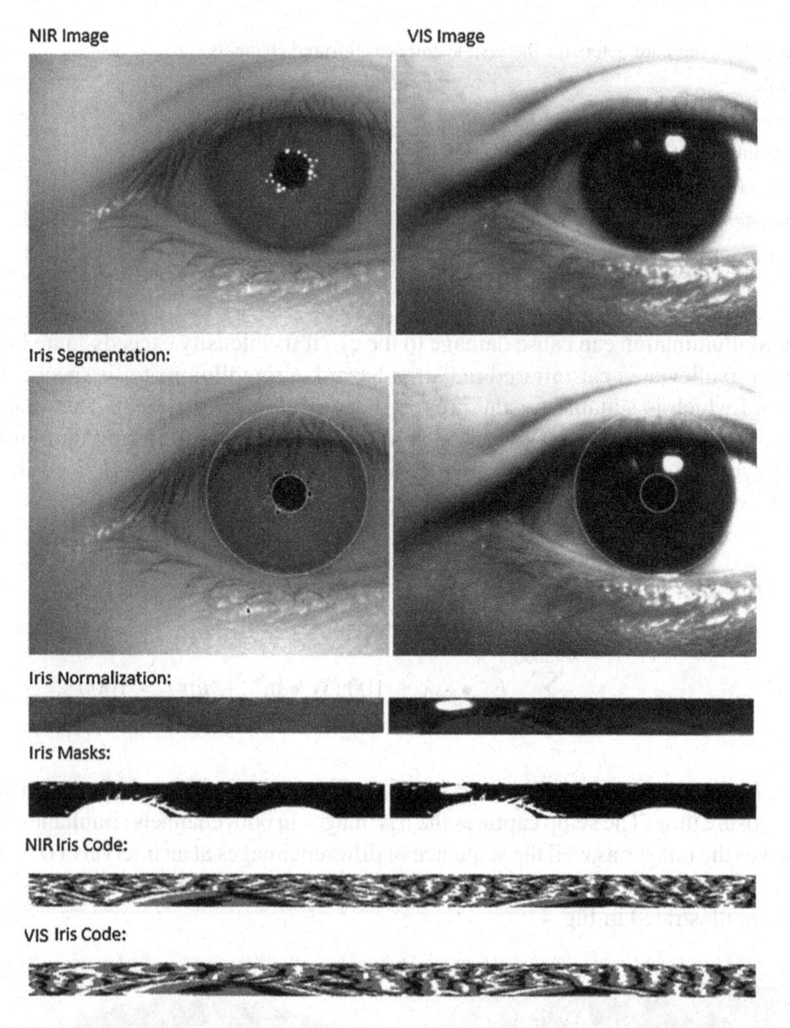


**Fig. 5** Preprocessing steps for cross-spectral iris recognition

# 3 Experiments and Results

Our key objective is to ascertain true cross-spectral matching capability from the simultaneously acquired iris under two popular spectrums. The matching accuracy for cross-spectral iris matching is ascertained from the equal error rate and the decidability index, in addition to the receiver operating curve. The decidability index ($d'$) can be defined as

$$d' = \frac{|\mu_A - \mu_I|}{\sqrt{\frac{1}{2}(\sigma_A^2 + \sigma_I^2)}} \tag{2}$$

where $\mu_A$ and $\mu_I$ are the means of authentic and impostor distance score distributions and $\sigma_A$ and $\sigma_I$ are the standard deviations of the two distributions. We develop a true cross-spectral iris image database using the proposed iris acquisition setup. The database consists of a total of 400 iris images ($10 \times 2 \times 2 \times 10$) that are acquired in 10 instances from 10 subjects of both eyes simultaneously in both near infrared and visible channels. The iris segmentation, normalization, and template extraction are performed for both NIR and VIS images (see Fig. 5) using the algorithm in [20]. A standard 1-D log-Gabor filter was convolved with the normalized iris image of size, $20 \times 240$, in order to create a binarized iris code of size, $20 \times 240 \times 2$. The normalized hamming distance measure is used to compute the dissimilarity between iris codes. The performance measures are shown in Table 3 for three different types of cross-comparisons, namely near infrared to near infrared (NIR-to-NIR), visible to visible (VIS-to-VIS), and infrared to visible (NIR-to-VIS) comparisons. The decidability index ($d'$) and equal error rate (EER) for cross-spectral iris comparisons (NIR-to-VIS) are 0.9253 and 34.019, respectively. From Figs. 6 and 7, it can be observed that even though the cross-comparisons in near infrared and visible channels are independently quite accurate, the performance of cross-spectral iris matching has significantly degraded. Therefore, there is need to develop large-scale cross-spectral iris database and advanced algorithms to accurately match such images.

**Table 3** Results of three cross-comparisons (NIR-to-NIR, VIS-to-VIS and NIR-to-VIS)

| Cross-comparison | $\mu_A$ | $\mu_I$ | $\sigma_A$ | $\sigma_I$ | $d'$ | EER |
|---|---|---|---|---|---|---|
| NIR-to-NIR | 0.2463 | 0.4590 | 0.0875 | 0.0136 | 3.3949 | 2.9447 |
| VIS-to-VIS | 0.2704 | 0.4428 | 0.0898 | 0.0200 | 2.6503 | 7.2337 |
| NIR-to-VIS (cross spectral) | 0.4400 | 0.4613 | 0.0298 | 0.0133 | 0.9253 | 34.019 |

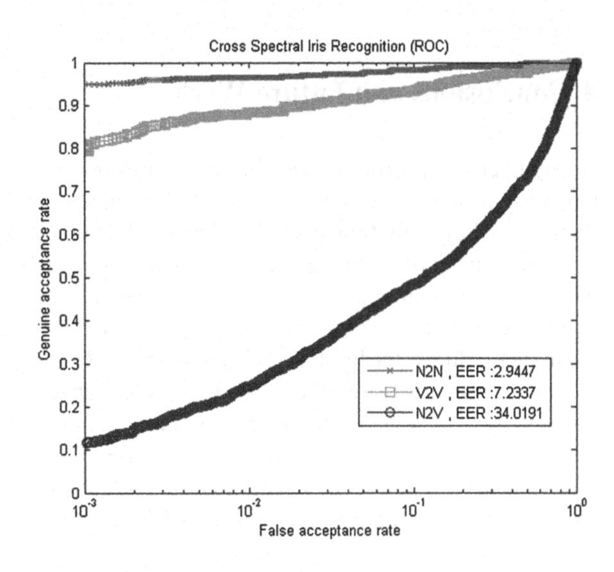


**Fig. 6** The ROC curves for cross-spectral iris recognition (N2N: NIR-to-NIR; V2V: VIS-to-VIS and N2V: NIR-to-VIS comparison

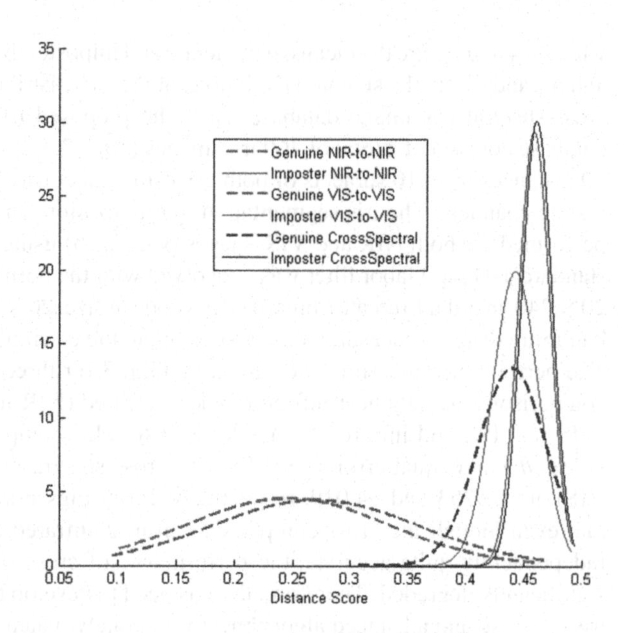


**Fig. 7** Normalized histograms plots of genuine and impostor distance scores for NIR-to-NIR, VIS-to-VIS and NIR-to-VIS comparisons

**Table 4** Performance degradation for cross-sensor iris matching: EER and DI on ND dataset

| Comparisons | Pillai et al. | Lihu Xiao et al. | Lihu Xiao et al. |
|---|---|---|---|
| | EER (%) [14] | EER (%) [15] | $d'$ (%) [15] |
| LG2200 | 6.06 | 1.84 | 4.1169 |
| LG4000 | 5.22 | 0.31 | 5.0545 |
| Cross-sensor | 7.19 | 3.38 | 3.6546 |
| Improved cross-sensor | 6.09 | 0.91 | 5.3070 |

## 4 Conclusions and Future Work

Accurate cross-spectral iris capability requires iris image database which is simultaneously acquired under two different spectrums. Lack of such database in public domain is a key limitation to advanced research in this area. This paper has described the development of a benchmark cross-spectral iris database in which the iris images were acquired simultaneously under near infrared and visible regions of electromagnetic spectrum. To the best of our knowledge, there is no standard cross-spectral iris database to address the limitations in cross-spectral iris recognition. The performance of cross-sensor iris recognition [14, 15] is summarized in Table 4 for the entire ND dataset. The extent of performance degradation for cross-sensor iris matching can be observed from this table. Considering our results, particularly on a smaller database, it is *reasonable to argue that cross-spectral iris matching seriously degrades the iris matching accuracy than the cross-sensor iris matching accuracy.*

Our observations from fully registered true cross-spectral iris matching illustrate serious degradation in matching accuracy. This can be explained by the fact that the information revealed under visible and near infrared spectrum is expected to be significantly different but have some correlation compared to those observed from different sensors operating under the same near infrared spectrum. The extent of such cross-spectral correlation and difference need to be carefully investigated in the future work. In this context, the cross-sensor framework in [14] can be explored on registered cross-spectral data. Development of such cross-spectral iris recognition using advanced machine learning algorithms is part of our ongoing research in this area.

**Acknowledgments** This work is supported by research grant from Hong Kong Research Grant Council grant no. PolyU/152068/14E.

## References

1. Global Biometrics Market (2014–2020): Market Forecast By Technologies, Applications, End Use, Regions and Countries. http://www.researchandmarkets.com/research/6lngtl/global_biometrics/
2. Daugman, John G.: Pattern analysis and machine intelligence. IEEE Trans. High Confid. Vis. Recognit. Pers. Test Stat. Indep. **15**(11), 1148–1161 (1993)
3. Bowyer, K.W., Hollingsworth, K., Flynn, P.J.: Image understanding for iris biometrics: a survey. Comput. Vis. Image Underst. Elsevier **110**(2), 281–307 (2008)
4. Proença, H.: Non-cooperative iris recognition: issues and trends. In: Proceedings of the EUSIPCO, pp. 1–5 (2011)
5. Imai, F.H.: Preliminary experiment for spectral reflectance estimation of human iris using a digital camera, Munsell Color Science Laboratory Technical Report (2002)
6. Proença, H., Alexandre, L.A.: UBIRIS: a noisy iris image database. In: Springer Proceedings of International Conference on Image Analysis and Processing, pp. 970–977 (2005)
7. Proença, H., Filipe, S., Santos, R., Oliveira, J.: Alexandre Luis: The ubiris. v2: a database of visible wavelength iris images captured on-the-move and at-a-distance. IEEE Trans. Pattern Anal. Mach. Intell. **32**(8), 1529–1535 (2010)
8. Dobes, M., Machala, L.: UPOL iris image database. http://www.phoenix.inf.upol.cz/iris/ (2004)
9. Tajbakhsh, N., Araabi, B.N., Soltanianzadeh, H.: Feature fusion as a practical solution toward noncooperative iris recognition. In: IEEE Proceedings of 11th International Conference on Information Fusion, pp. 1–7 (2008)
10. Tan, C.-W., Kumar, A.: Unified framework for automated iris segmentation using distantly acquired face images. IEEE Trans. Image Process. **21**(9), 4068–4079 (2012)
11. Ross, A., Pasula, R., Hornak, L.: Exploring multispectral iris recognition beyond 900 nm. In: IEEE Proceedings of 3rd International Conference on Biometrics: Theory, Applications, and Systems, pp. 1–8 (2009)
12. Sharma, A., Verma, S., Vatsa, M., Singh, R.: On cross spectral periocular recognition. In: IEEE Proceedings of 21st International Conference on Image Processing, pp. 5007–5011 (2014)
13. Zuo, J., Nicolo, F., Schmid, N.A.: Cross spectral iris matching based on predictive image mapping. In: IEEE Proceedings of 4th International Conference on Biometrics: Theory, Applications, and Systems, pp. 1–5 (2010)
14. Pillai, Jaishanker K., Puertas, Maria, Chellappa, Rama: Cross-sensor iris recognition through Kernal learning. IEEE Trans. Pattern Anal. Mach. Intell. **36**(1), 73–85 (2014)

15. Xiao, L., Sun, Z., He, R., Tan, T.: Coupled feature selection for cross-sensor iris recognition. In: IEEE Proceedings of 6th International Conference on Biometrics: Theory, Applications, and Systems, pp. 1–6 (2013)
16. Kulis, B., Sustik, M., Dhillon, I.: Learning low-rank kernel matrices. In: ACM Proceedings of the 23rd International Conference on Machine Learning, pp. 505–512 (2006)
17. Davis, J.V., Kulis, B., Jain, P., Sra, S., Dhillon, I.S.: Information-theoretic metric learning. In: ACM Proceedings of the 24th International Conference on Machine Learning, pp. 209–216 (2007)
18. JAI, Multi-spectral imaging, http://www.jai.com/en/products/
19. Intersil, Eye safety for proximity sensing using infrared light-emitting diodes, http://www.intersil.com/content/dam/Intersil/documents/an17/an1737.pdf
20. Masek, Libor and Kovesi, Peter, Matlab source code for a biometric identification system based on iris patterns, The School of Computer Science and Software Engineering, The University of Western Australia, 2(4), (2003)

# Fast 3D Salient Region Detection in Medical Images Using GPUs

**Rahul Thota, Sharan Vaswani, Amit Kale and Nagavijayalakshmi Vydyanathan**

**Abstract** Automated detection of visually salient regions is an active area of research in computer vision. Salient regions can serve as inputs for object detectors as well as inputs for region-based registration algorithms. In this paper, we consider the problem of speeding up computationally intensive bottom-up salient region detection in 3D medical volumes. The method uses the Kadir–Brady formulation of saliency. We show that in the vicinity of a salient region, entropy is a monotonically increasing function of the degree of overlap of a candidate window with the salient region. This allows us to initialize a sparse seed point grid as the set of tentative salient region centers and iteratively converge to the local entropy maxima, thereby reducing the computation complexity compared to the Kadir–Brady approach of performing this computation at every point in the image. We propose two different approaches for achieving this. The first approach involves evaluating entropy in the four quadrants around the seed point and iteratively moving in the direction that increases entropy. The second approach we propose makes use of mean shift tracking framework to affect entropy maximizing moves. Specifically, we propose the use of uniform pmf as the target distribution to seek high entropy regions. We demonstrate the use of our algorithm on medical volumes for left ventricle detection in PET images and tumor localization in brain MR sequences.

R. Thota
Target, Bangalore, India
URL: http://www.target.com

S. Vaswani
Department of Computer Science, University of British Columbia,
Vancouver, Canada

A. Kale (✉) · N. Vydyanathan
Imaging and Computer Vision Group, Siemens Corporate
Research and Technology, Prestige Alecto Building,
Electronics City, Bangalore 560100, India
e-mail: kale.amit@siemens.com



R. Singh et al. (eds.), *Machine Intelligence and Signal Processing*,
Advances in Intelligent Systems and Computing 390,
DOI 10.1007/978-81-322-2625-3_2

# 1 Introduction

Images contain great amounts of information at multiple frequencies, scales, and spatial locations. Perceiving this information in its entirety is a hard task. The human brain has evolved neurological processes that selectively focus attention [1] at one salient frequency, scale, or image location at a time. In order to enable processing of large amounts of data, it is natural to automate these neurological processes. This has led to the development of automated visual saliency detection algorithms that estimate salient features in images [2, 3]. Frintrop et al. in [4] showed the higher repeatability and discriminative power of salient regions as compared to ensemble-based detections, leading to more accurate matching across different scenes for registration or 3D scene estimation. Object detection and recognition accuracies have been improved by applying saliency filter as the front end followed by specific descriptors like SIFT [5] trained on the object class.

Visual saliency, can be thought of as a combination of bottom-up and top-down attention [6, 7]. Top-down uses the prior knowledge, models, and abstractions [8]. Judd et al. [9] used a combination of low-level local features and semantic information from high-level detectors for faces and people. Such top-down approaches try to predict the way humans perceive the visual world. It is difficult to translate such an approach to domains like medical image analysis where human attention is task-dependent. Hence, we primarily concentrate on bottom-up saliency, which predominantly depends on the conspicuity emerging from contrasting/distinguishing local image features. For instance, Mahadevan and Vasconselos [10] proposed an architecture where saliency is proportional to the discriminability of a set of features that classify center from surround, at that location. They set the size of the center window to a value comparable to the size of the display items and the surround window six times the size. This ratio is motivated from the neurological evidences on natural images. However, such an assumption does not hold in medical images where the lesions and organs can take diverse range of sizes. Itti and Koch [2] deal with this problem by estimating the size of salient region by applying difference of Gaussians on the image pyramid using different combinations of standard deviations that capture different center–surround ratios. This adds an extra dimension of complexity leading to high compute expense. Le Ngo et al. [11] estimated the conditional entropy of the center given its surrounds using kd-tree to reduce computation. This method assumes fixed size windows for defining center and surround region making it intractable for medical imaging where anatomies could be of various sizes. Also their method is not scalable for finding salient regions in three-dimensional volumes. Considering the above-mentioned shortcomings, we adopt the saliency definition proposed by Kadir and Brady [12]. Specifically, their framework considers different-sized neighborhoods of every point in an image. The maximum of the product of the local entropy and the rate of the pdf as a function of scale of the neighborhood is then computed. If this maxima exceeds a pre-decided threshold the point is designated as a salient point. Subsequently, Expectation–Maximization (EM) based clustering is employed to coalesce nearby detections. These steps of computing entropy and differential pdf at every point in the image at multiple scales are computationally intensive, especially when considering 3D volumetric imagery common in medical imaging.

In this paper, we propose an approach for fast salient region detection in 3D imagery based on the Kadir–Brady approach. We show that in the vicinity of a salient region, entropy is a monotonically increasing function of the degree of overlap of a candidate window with the salient region. This allows us to initialize a sparse seed point grid as the set of tentative salient region centers and iteratively converge to the local entropy maxima. This reduces the computation considerably compared to the Kadir–Brady approach of performing this computation at every point in the image. We propose two different approaches for achieving this. The first approach involves evaluating entropy in the four quadrants around the seed point and iteratively moving in the direction that increases entropy. In particular, the effective displacement is calculated as the summation of four quadrant displacements weighted by corresponding normalized entropies. The second approach we propose makes use of pixel-level information in a mean shift tracking framework to effect entropy maximizing moves. Specifically, we propose the use of uniform pmf as the target distribution to seek high entropy regions. We also extend this Saliency shift algorithm to capture 3D salient regions in medical volumes by estimating orientation using 3D extension of ABM-SOD algorithm [13]. We develop an optimized GPU implementation of the saliency seek algorithm to enable and accelerate the detection of salient regions. We demonstrate results for left ventricle detection in PET and for the tumor map in brain MR sequence.

# 2 Technical Details

## 2.1 Motivation

As discussed before, the Kadir–Brady approach considers different-sized neighborhoods of every point in an image. The maximum of the product of the local entropy and the rate of the pdf as a function of scale of the neighborhood is then computed. If this maxima exceeds a pre-decided threshold the point is designated as a salient point. Subsequently, Expectation–Maximization (EM) based clustering is employed to coalesce nearby detections. The key problem with this approach is the need to perform this computation at every point in the image which as we show can be quite unnecessary. To motivate this, consider a toy example consisting of a single salient region as shown in the Fig. 1.

We assume that the intensity distribution $f$ in the salient region follows a uniform distribution and the background distribution $g$ follows a delta distribution. Specifically, $f \sim U[0, M]$ and $g \sim \delta(k - l)$. The choice of these distributions captures the fact that salient regions have higher entropy as compared to their background in an exaggerated sense. Now consider a candidate window with an overlap fraction $\alpha$

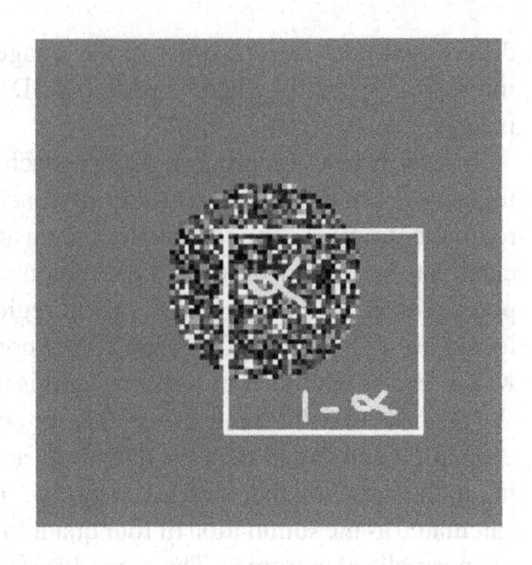


**Fig. 1** Toy example to demonstrate monotonicity of entropy against percentage overlap with the salient region. The figure shows a target object with a uniform range of intensities in a homogeneous background region. The candidate window partially overlaps with the target with the fraction of overlap given by $\alpha$

with the salient region. It is easy to see that the intensity distribution of this window will be a mixture distribution $\alpha f + (1-\alpha)g$. The entropy of this mixture distribution is given by:

$$H(\alpha) = -\sum_{k=0}^{M} (\alpha f + (1-\alpha)g) \log (\alpha f + (1-\alpha)g) \tag{1}$$

Substituting the distributions for $f$ and $g$ from above, the entropy evaluates to

$$H(\alpha) = -(\frac{\alpha}{M} + 1 - \alpha) \log (\frac{\alpha}{M} + 1 - \alpha) - \alpha \frac{M-1}{M} \log \frac{\alpha}{M} \tag{2}$$

The rate of change of this entropy as a function of $\alpha$ is given by Eq. 3

$$\frac{d}{d\alpha}(H(\alpha)) = \frac{M-1}{M} \log \frac{\alpha + M(1-\alpha)}{\alpha} \tag{3}$$

Note that for $\alpha = 0$ $H(\alpha) = 0$ and for $\alpha = 1$, $H(\alpha) = log(M)$. Also, it can be seen that the above expression in Eq. 3 for the derivative of the entropy is positive, for $\alpha$ ranging between 0 and 1. This shows that for the simple toy example that we have considered, that entropy increases monotonically as a function of overlap percentage in the vicinity of the salient object. In turn, this relationship suggests that, in the neighborhood of a salient region, an iterative algorithm that increases entropy of the enclosed region, will increase the overlap percentage. This would result in moving the candidate window closer to the salient object, obviating the need to perform an exhaustive entropy computation at every point.

In the following sections, we describe two approaches which use the above idea to detect salient regions starting from a sparse set of initial seed points. The first approach involves evaluating entropy in the four quadrants centered around the seed point and iteratively moving in the direction that increases entropy. In particular, the effective displacement is calculated as the summation of four quadrant displacements weighted by corresponding normalized entropies. The second approach we propose makes use of mean shift tracking framework to effect entropy maximizing moves. Specifically, we propose the use of uniform pmf as the target distribution to seek high entropy regions. We would like to note here that while the above assumptions made regarding the distributions of the salient region and its background do not strictly hold for real-world problems, the proposed entropy maximizing algorithms, nevertheless succeed in converging to the salient regions.

## 2.2 Quadrant Method

In this algorithm, we initialize a uniformly sampled grid of points on the image. In order to capture all the entropy maxima, we assume that each salient region has at least one grid point within its vicinity. Let us denote the locations of these initial grid points with $\bar{P}_i^0$ where the subscript represents the index of the $i$th point and the superscript denotes the iteration. Each of the points in the grid represents the tentative center of a salient region for the corresponding iteration. At every step, direction of increasing entropy has to be calculated. This is achieved by dividing the neighborhood of the point considered into four quadrants as shown in the figure below. In each of the quadrants we take windows $W_{jk}(P_i^t)$ of all scales in a range varying over $k$ and compute respective entropies. For each quadrant, the window at the scale that gives maximum entropy is selected. The effective shift vector $\delta \bar{ed_i}$ is calculated as the summation of the displacement vectors corresponding to the optimal scales weighted by their normalized entropies. For more details, see Algorithm 1 and Fig. 2a.

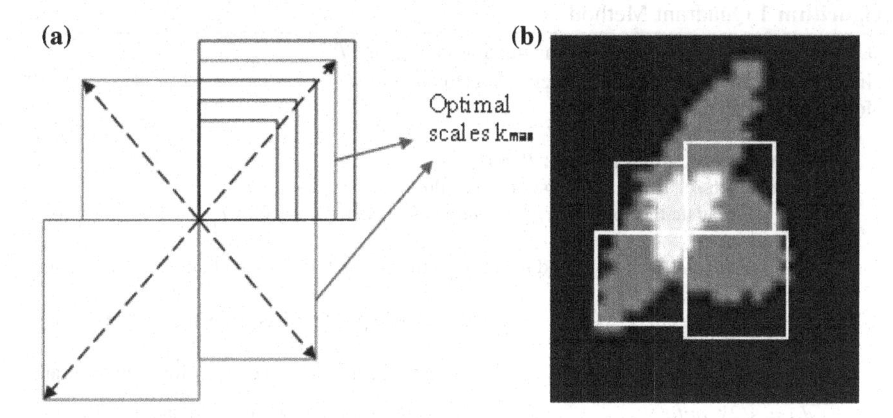


**Fig. 2** **a** The figure shows the 4 quadrants for a seed point. It shows the range of scales for entropy calculation for one of the quadrants. The optimum scales for each of the quadrants are shown in *red* with *dotted lines* representing the displacement vectors. **b** A sample result obtained using the quadrant method on a synthetic image figure. As can be seen, the four quadrants localize the target salient object.

Once entropy maxima are found we calculate the change in pdf about the optimal scale. Salient regions are obtained by filtering out the high entropy noise by applying a lower bound on the pdf difference. A sample result of this approach is shown in Fig. 2b. The quadrant approach considers variable-sized neighborhoods of a test point and makes an entropy-weighted move toward the salient region. The entropy weights are coarse, however, in the sense that they are computed for each quadrant rather than pixel level. In the next section, we propose a more fine-grained approach to make more precise shifts to higher entropy regions. While the idea of using quadrants is feasible for 2D problems, its extension to 3D volumes is computationally intensive. Furthermore, it cannot deal with anisotropic salient regions as described in the next section.

### 2.3 Saliency Shift

Mean shift tracking [14] is a popular approach for nonrigid object tracking based on maximizing the Bhattacharyya coefficient between histogram of the target and the candidate window in successive frames. This maximization is achieved by an iterative procedure which involves computing a weight map over the candidate window that reflects the likelihood of a pixel belonging to the target histogram. A shift vector pointing to the centroid of the weight map is then computed and the procedure is repeated till convergence. We adapt this concept for maximizing entropy. One problem with this is that unlike the visual tracking problem where the target histogram refers to a fixed template, we do not have a specific target histogram to work with for the entropy problem. This can be addressed as follows: We note that among all

---

**Algorithm 1** Quadrant Method

---

Uniformly sampling the image with an initial grid of points $P_i^0$
Initialize magnitude effective displacement vector $\delta\bar{ed}_i$ to $\eta$
**for each** point $P_i$ **do**
  **while** $\|\delta\bar{ed}_i\| \leq \eta$ **do**
    **for each** quadrant j centered at $P_i$ **do**
      **for each** scale k in the range of scales **do**
        Calculate the entropy $\epsilon_{ijk}$ in window $W_{jk}$ centered at point $P_i^t$
      **end for**
      $\epsilon_{ij} = max_k \epsilon_{ijk}$ ▷ Find the maximum entropy for each quadrant across the range of scales
      $k_{opt_{ij}} = argmax_k \epsilon_{ijk}$ ▷ Find the corresponding scale maximizing the entropy
    **end for**
    $n\epsilon_{ij} = \frac{\epsilon_{ij}}{\sum_j \epsilon_{ij}}$ ▷ Find the normalized entropy for each quadrant
    $\bar{d}_{ij} = \sqrt{2}k\_opt_{ij}$ ▷ Displacement in each quadrant direction
    $\delta\bar{ed}_i = \sum_j n\epsilon_{ij}\bar{d}_{ij}$ ▷ Calculate effective displacement for the point
    $t = t + 1$
    $\bar{P}_i^{t+1} = \bar{P}_i^t + \delta\bar{ed}_i$ ▷ Change location of point
  **end while**
**end for**

---

discrete distributions, the uniform distribution has highest entropy. In order to adapt the mean shift tracking procedure to seek entropy maxima, a simple solution is to use the uniform distribution as the target histogram. We now describe the procedure in detail. A sparse set of seed points is distributed throughout the image. A search window $W_x$ is centered around each candidate point $x$ and candidate feature distribution $p$ of the the window is calculated by aggregating the weighted kernel responses over all the pixels in it.

$$p_b(W_x) = C_H \sum_{s \in W} |H|^{-1/2} K(d(x, s)) \delta[\beta(s) - b] \tag{4}$$

where $b$ is the histogram bin index, $\beta(s)$ is the bin number in which pixel $s$ lies, $C_H$ is the normalization constant, $H$ denotes the bandwidth matrix, $\delta$ is the discrete Kronecker delta function, and $d$ is the Euclidean distance. As mentioned above, to move toward higher entropy regions, we define the target distribution to be uniform $q \sim U[0, M]$. The algorithm estimates the shift by maximizing the similarity between the candidate distribution and uniform pdf, measured in terms of the Bhattacharyya coefficient given by $\rho(x) = \sum_{b=1}^{M} \sqrt{p_b(W_x) q_b}$. Each pixel in the candidate window is assigned weights given by Eq. 5

$$w(s) = \sum_{1}^{M} \sqrt{q_b / p_b(x)} \delta[\beta(s) - b] \tag{5}$$

which in effect assigns higher importance to the candidate pixels belonging to the target distribution. At each iteration, the next position $x$ of the seed point is calculated by a kernel $K$ weighted mean of the candidate window pixels.

$$\mathbf{x} = \frac{\sum_{s \in S} -K'_H(x - s) w(s) \mathbf{s}}{\sum_{s \in S} -K'_H(x - s) w(s)} \tag{6}$$

The algorithm increasing the Bhattacharyya coefficient with the uniform distribution effectively moves the candidate in a higher entropy direction till the maximum is achieved. All maxima with entropy values above a certain predefined threshold are selected and the change in pdf is calculated at the specified scale. Regions with high rate of change of pdf are considered to be salient at that scale. We use above-mentioned framework for identifying 3D salient regions in medical volumes. Specifically, we use cuboid search windows that are initially distributed randomly throughout the medical volume. Each of these windows has a scale parameter generated uniformly in a range constrained by the size of the volume. The size of the window does not change across iterations, i.e., the bandwidth matrix $H$ remains the same throughout the mean shift procedure. We use the identity function as our kernel $K(x) = x$. This approach worked well for isotropic salient regions as well as for anisotropic regions aligned with the coordinate axes. However for salient regions which are anisotropic and oblique, the cuboids lack the flexibility to precisely encapsulate the salient region resulting in a higher fraction of the background pixels contributing to the candidate histogram thus reducing its entropy and resulting in a missed detection as can be seen in Fig. 3.

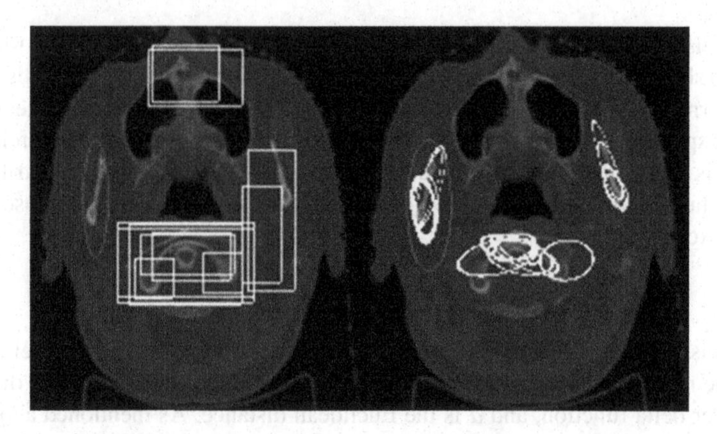

**Fig. 3** The image on the *left* shows a 2D projection of the output for the saliency detection algorithm using the conventional mean shift framework, whereas the *right* image shows the same volume slice using the ABMSOD framework. The cheek bones are not detected using conventional mean shift because of the anisotropy

To address this problem, we use Adaptive Bandwidth Mean shift for Object Detection (ABMSOD) algorithm [13]. ABMSOD is a mean shift-based iterative algorithm used for 2D object detection in computer vision. It simultaneously estimates the position, scale, and orientation of the target object. The initialization step involves randomly scattering elliptical candidate windows with varying sizes throughout the volume. The iterative step of the algorithm consists of two parts. In the first part, new estimate of the candidate location is calculated using the conventional mean shift framework. In the second part, we estimate the scale and orientation parameters that are encoded within the bandwidth matrix of the candidate window. For details of the implementation see Algorithm 2. The optimal bandwidth corresponds to the scale and orientation that maximizes the Bhattacharyya coefficient at the current position. Chen et al. [15] derive the expression for estimating the optimum bandwidth matrix $H$.

$$H = \frac{\sum_{s \in S}(x - s)(x - s)^T w(s)}{\sum_{s \in S} w(s)} \tag{7}$$

Vaswani et al. [16] validate the expression Eq. 7 for higher dimensions and use the ABMSOD framework for localizing 3D structures in medical volumes. For this, the feature histogram of the anatomy to be localized is used as the target distribution. We adopt the same framework for maximizing the entropy using uniform pdf as the target distribution. However, the algorithm described in Algorithm 2 is computationally expensive and is amenable to parallelization using GPUs. The scheme used for parallelization of our algorithm is described in the next subsection.

**Algorithm 2** Adaptive Bandwidth Mean shift

HQ is the target histogram
**for each** candidate ellipsoid $W_i$ centered at $P_i$ with a bandwidth matrix $H_i$ **do** ▷ Perform iterative search
$iter\_cnt \leftarrow 1$
$bhatcf \leftarrow 0$ ▷ Initialize Bhattacharyya coefficient
$max\_bhatcf \leftarrow 0$ ▷ Maximum Bhattacharaya coefficient across all iterations
$delta\_bhat \leftarrow THRESHOLD$
$x_{opt} \leftarrow P_i$
$H_{opt} \leftarrow H_i$
**while** $(delta\_bhatcf \leq THRESHOLD$ and $iter\_cnt < MAX\_ITERATIONS)$ **do** ▷ Termination criteria for the meanshift search
**for each** voxel $v \in W_i$ **do**
$HP(b_v) \leftarrow HP(\beta(v)) + c \exp \frac{-d_v}{2}$ ▷ $HP$ is the candidate histogram local to each thread, $\beta(v)$ is the bin index for voxel $v$, and $d_v$ is the distance of voxel $v$ from the center of the ellipsoid
**end for**
**for each** voxel $v \in W_i$ **do**
$w_v \leftarrow \sqrt{\frac{HQ(\beta(v))}{HP\beta(v)}}$ ▷ Weight is computed for each voxel as the ratio of the bin heights of the candidate and target histograms.
$\delta_{xv} \leftarrow -K'H(v)w_v s_v$ ▷ $\delta_{xv}$ denotes the contribution of the voxel $v$ having position $s_v$ to the change in the position of the ellipsoid
$x_{new} \leftarrow \frac{\sum_{\forall v} \delta_{xv}}{\sum_{\forall v} -K'H(v)w_v}$ ▷ $x_{new}$ is the new position of candidate ellipsoid and is computed according to Eq. 6
**for each** voxel $v \in W_i$ **do**
**Repeat** Steps 11 to Step 14 for the candidate ellipsoid $W_i$ centered at $x_{new}$
$\delta_{Hv} \leftarrow (x_{new} - s_v)(x_{new} - s_v)^T w_v$ ▷ $\delta_{Hv}$ denotes contribution of a voxel to the H matrix
**end for**
$H_{new} \leftarrow \frac{\sum_{\forall v} \delta_{Hv}}{\sum_{\forall v} w_v}$ ▷ $H_{new}$ is the optimum H matrix at the new position of the candidate ellipsoid and is calculated using Eq. 7
$bhatcf \leftarrow \sqrt{(HQ)(HP)}$
$delta\_bhatcf \leftarrow max\_bhatcf - bhatcf$
**if** $bhatcf > max\_bhatcf$ **then**
$max\_bhatcf \leftarrow bhatcf$
$x_{opt} \leftarrow x_{new}$
$H_{opt} \leftarrow H_{new}$
**end if**
$iter\_cnt \leftarrow inter\_cnt + 1$
**end for**
**end while**
$P_{i\_opt} \leftarrow x_{new}$
$H_{i\_opt} \leftarrow H_{new}$
**end for**

## 2.4 Parallelization Using GPUs

In this subsection, we describe the scheme used to parallelize the ABMSOD algorithm on a GPU. A similar scheme is used to accelerate the conventional mean shift algorithm. The GPU is a data-parallel computing device consisting of a set of multiprocessing units (SM), each of which is a set of SIMD (single instruction multiple data) processing cores. Each SM has a fixed number of registers and a fast on-chip memory that is shared among its SIMD cores. The different SMs share a slower off-chip device memory. Constant memory and texture memory are read-only regions of the device memory and accesses to these regions are cached. Local and global memory refers to read–write regions of the device memory and its accesses are not cached. In the CUDA context, the GPU is called device, whereas the CPU is called host. Kernel refers to an application that is executed on the GPU. A CUDA kernel is launched on the GPU as a grid of thread blocks. A thread block contains a fixed number of threads. A thread block is executed on one of the multiprocessors and multiple thread blocks can be run on the same multiprocessor (For details on GPU architecture see [17]). The independent exploration of different search paths originating from each initial random point is distributed amongst thread blocks. In addition, using the finer level of parallelism offered by threads within a thread block, we further parallelize operations within each search iteration. We make the threads in a thread block handle computations for a subset of the voxels from the window. Computations such as the application of the kernel function, construction of the candidate histogram, weight assignment to the voxels, H matrix computation, etc. are all done in parallel where each thread is responsible for a set of voxels. Summation of values across threads is performed through parallel reduction. To reduce the synchronization operations among threads during histogram computation, we allow each thread to construct a local histogram of the voxels handled by that thread. These histograms are stored in shared memory for fast access. After all local histograms are constructed, the histograms are binwise aggregated by the threads in parallel to form the global candidate histogram. The volume is stored in a 3D texture and the access to the volume is ensured to be in a way that maximizes spatial locality and efficiently utilizes the texture cache. Constant variables that target the histogram are stored in constant memory to utilize the constant cache. Data shared by threads within a block like the local histograms, etc. are stored in shared memory for fast retrieval through the broadcast mechanism supported by CUDA. Enough number of threads are launched to keep all the cores of each streaming multiprocessor (SM) busy. We try and maximize the occupancy for each SM. The occupancy is, however, limited by the amount of shared memory and the number of registers available per SM. We also ensure that there is no register spilling and uncoalesced global memory accesses.

# 3 Experiments

We consider the usage of fast salient region detection in order to speed up medical workflows. Specifically, we demonstrate the efficacy of our algorithm in estimating the location and size of left ventricle in myocardial perfusion imaging and quantifying brain tumor in MR images acquired by Fluid Attenuated Inversion Recovery Pulse Sequence (FLAIR).

## *3.1 LV Detection in Myocardial Perfusion Imaging*

Myocardial perfusion imaging is a nuclear medicine procedure that illustrates the function of the heart muscle (myocardium). The patient is typically administered FDG—fluorodeoxyglucose—which has radioactive isotope fluorine that emits imagable positrons. This technique captures the functional information of the body as against structural information from CT because the glucose part in the radiopharmaceutical rushes to regions in the body such as the myocardium which have high metabolic activity. Physicians require the myocardium dataset in a standard orientation and scale. Current techniques that estimate cardiac orientation [18] need a good initial mask around the myocardium otherwise, liver and other nearby organs having high uptake contribute to a biased estimate of the orientation and size.

Detection of Left ventricle in a medical volume of size $128 \times 128 \times 34$ using object appearance-based classifier at all possible locations and scales [18] is extremely compute-intensive. Myocardium in PET has increased uptake value because of high metabolic activity. One way to solve the problem of high computation is using this high uptake principle. Setting a high SUV threshold gives a rough initial estimate of the heart location. But we also observed large number of false positives due to noise, liver, and other organs Table 1. Also, such a simple threshold does not give any estimate of the size in each dimension. Exploring alternatives, we observed that left ventricular region can be considered salient because of the high entropy and uniqueness of its pixel intensity distribution compared to its immediate surroundings. In order to leverage that property, we propose to use 3D saliency seek as a preprocessing step to identify tentative candidates having left ventricle. The candidates would then be further processed for location and cardiac orientation estimation. This two-step approach eliminates false positives, effectively improving the accuracy and reducing

**Table 1** Comparison of saliency seek versus intensity-based thresholding

| Method | Recall Rate (for 20 detections) | Final precision (using Hu moments) |
|---|---|---|
| Threshold | $15/32 = 46\,\%$ | N.A* |
| Saliency Seek | $29/32 = 91\,\%$ | 25/32 |

*The intensity thresholding method does not give scale information so it cannot be used for further Hu-based detection

compute time. We apply the saliency seek algorithm on an initial sparse seed point grid distributed uniformly across the volume at multiple initial scales. Once these points converge to local saliency maxima, we pick the top 20 having high pdf difference w.r.t surrounding. We observed that one of the top 20 detections is a true positive in 29 of the 32 volumes dataset giving us a recall rate of 91 %.

In order to increase the precision, i.e., to identify the true positives among the tentative candidates, we used Hu moments [19]. These central moments have been carefully designed to be invariant to translation, rotation, scale and so they serve as a good choice for describing local appearance. We evaluated the set of seven invariant Hu moments on the center slice of a heart template forming a seven-dimensional training feature vector. For each test volume, we then compute the seven-dimensional Hu-moment test vectors on five slices about the central slice for each detection. The detection which minimizes the average Euclidean distance with the template across the five slices is chosen as the most accurate estimate for that volume. We used a test dataset of 32 PET volumes having a combination of "rest" and "stress" acquisitions, each of size $128 \times 128 \times 34$. The number of initial seed points was chosen to be 400. Also the average number of iterations for convergence was 10. We are able to detect the left ventricle by minimizing the distance in 25 volumes accurately. The average Jaccard index between ground truths and successful detections is as high as 41.36 %. See Fig. 4. Since we have to search for the ventricle structure throughout the PET

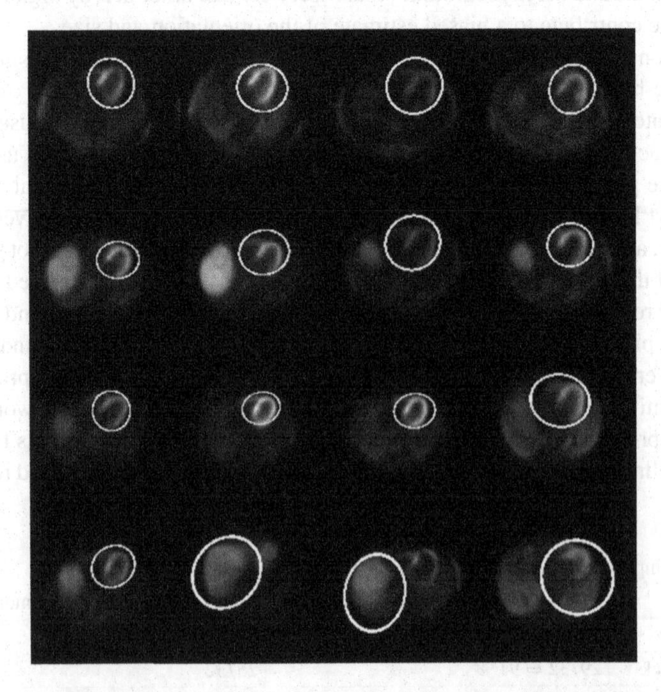

**Fig. 4** Results for LV detection on PET volumes. The figures show 2D axial slices for 16 PET volumes along with the final ellipsoids. The LV is correctly detected in all but two cases

volume, the algorithm is computationally intensive. To detect the left ventricle in a reasonable amount of time, we use the Nvidia Tesla C2050 GPU as an accelerator to speed up the saliency seek algorithm. This GPU has 14 multiprocessors each having 32 cuda cores, resulting in a total of 448 cuda cores. The cores are clocked at 1.15 GHz. Each multiprocessor has 48 KB of shared memory and 32 k registers. The GPU device has 3 GB of device memory. With the parallelization scheme as described in the previous section, the saliency seek algorithm is able to detect the left ventricle in 4.1 s which is 10 times faster as compared to a sequential implementation.

### 3.2 *Brain Tumor Quantification in MR*

In this section, we introduce the concept of user-defined saliency and demonstrate its accuracy in localizing brain tumor in MR volumes. In brain MRI, the accurate location of tumor and edema is essential for minimizing the damage to healthy tissue. In Fluid Attenuated Inversion Recovery (FLAIR) sequences, even subtle lesions stand out against attenuated csf fluids making this modality conducive for tumor detection, i.e., lesions can be distinguished based on intensity values alone and can be considered salient at the right window-level setting. Medical images like CT, MR, and PET have high dynamic ranges and different tissue types in the body lie in mutually exclusive intensity ranges [20, 21]. For example in CT, lung tissue ranges between $-400$ and $-600$ hounsfield units, fat tissue between $-60$ and $-100$, soft tissue lies between 40 and 80 HU, and bone from 400 to 1000 [22]. In such cases, the range of intensities defined by the window-level setting to be used is determined by anatomy of interest and the application at hand. We consulted a few clinical experts and found out the specific window-level setting used in FLAIR sequences to illustrate malignant lesions. Entropy evaluated over complete dynamic range of MR gave us unwanted regions straddling bone–tissue and air–tissue boundaries. We noticed much more relevant entropy values when the entropy is computed on the constrained intensity range coming from tumor specific window level. This new entropy quantifies the variation in the distribution of pixels falling in the relevant window level (Fig. 5).

We use the proposed saliency seek algorithm to locate tumors present in such MR volumes. However, we incorporated the new entropy evaluation as discussed above, using only a constrained intensity range. We found empirical evidence showing this idea of application specific saliency improving the detection accuracy significantly. We present the preliminary results of our evaluation of this concept on MR datasets. Brain tumor image data used in this work were obtained from [23]. The challenge database [23] contains fully anonymized images with manually labeled tumor ground truth. This dataset consists volumes of size $256 \times 256 \times 176$. We used the saliency seek algorithm in 15 such volumes and were able to successfully localize the tumor in all the 15 volumes with an average Jaccard index of 31.66 %.

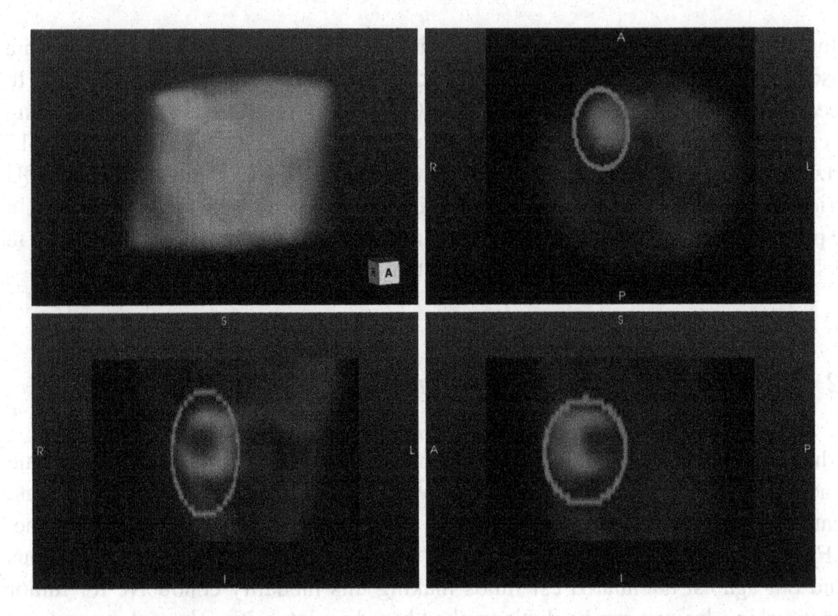


**Fig. 5** The four views (from *top* to *bottom*, *left* to *right*) represent the 3D, axial, coronal, and sagittal views of the PET volumes with LV ventricle detected by the final ellipsoid

In order to compare our results with the state of the art, we chose to apply a modified version of the Itti–Koch approach on axial 2D projections of the brain MR images. The modified algorithm considers only pixels in the constrained intensity range in constructing the saliency maps. In Fig. 6 the first row shows 2D slices from 3 MR volumes with a tumor. The second row isolates the tumor in each of the volume. The third row consists of the saliency maps obtained using the modified Itti–Koch algorithm. The fourth row consists of the set of detections obtained from the saliency seek algorithm. As can be seen from the figure, the Itti–Koch algorithm is not able to identify the tumor as a salient region in all the cases, whereas saliency seek employing the metrics of entropy and pdf difference is successfully able to localize the tumor. Since saliency seek in this case consists of searching for the tumor from numerous seed points scattered in a large-sized MR volume, the acceleration due to GPUs becomes important for detection in a reasonable amount of time. The number of initial seed points in this case was 700 with the average number of iterations 12. We use the Tesla C2050 GPU as described in Sect. 3.1 and are able to successfully localize the tumor in 7.8 s. We obtain a speedup of around 80x over the sequential CPU implementation.

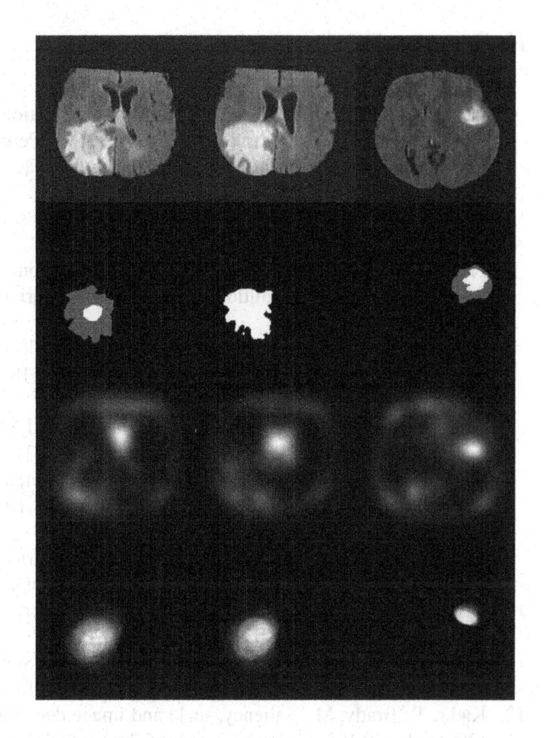

**Fig. 6** The *first row* shows 2D slices from 3 MR volumes with a tumor. The *second row* isolates the tumor in each of the volume. *Third row* shows the output of modified Itti–Koch algorithm. The *fourth row* consists of the set of detections obtained from the saliency seek algorithm

## 4 Conclusion

In this paper, we showed that in the vicinity of a salient region, entropy is a monotonically increasing function of the degree of overlap of a candidate window with the salient region and proposed two iterative approaches to locate salient regions from a sparse grid of seed points. The first used a four-quadrant approach to find entropy maximizing moves. The second used mean shift tracking framework using a uniform target distribution in mean shift iterations for seeking high entropy regions. We developed an efficient GPU implementation of the proposed algorithm for quickly detecting salient regions in 3D and showed promising results for Myocardium detection in PET volumes and tumor quantification in brain MR sequences. The framework can be easily extended for visual tracking by using the converged salient regions from previous frames. Also incorporating target specific feature information within the saliency shift iterations would be another interesting extension.

## References

1. Frintrop, S., Rome, E., Christensen, H.I.: Computational visual attention systems and their cognitive foundations: a survey. ACM Trans. Appl. Percept. (TAP) **7**, 6 (2010)
2. Itti, L., Koch, C.: Computational modelling of visual attention. Nat. Rev. Neurosci. **2**, 194–203 (2001)
3. Itti, L., Koch, C.: Feature combination strategies for saliency-based visual attention systems. J. Electron. Imaging **10**, 161–169 (2001)
4. Frintrop, S.: The High Repeatability of Salient Regions. http://hal.inria.fr/inria-00325799
5. Lowe, D.G.: Object recognition from local scale-invariant features. In: ICCV, vol. 2, pp. 1150–1157. IEEE (1999)
6. Desimone, R., Duncan, J.: Neural mechanisms of selective visual attention. Annu. Rev. Neurosci. **18**, 193–222 (1995). doi:10.1146/annurev.ne.18.030195.001205, http://dx.doi.org/10.1146/annurev.ne.18.030195.001205
7. Oliva, A., Torralba, A., Castelhano, M.S., Henderson, J.M.: Top-down control of visual attention in object detection. In: Proceedings of the ICIP (2003)
8. Corbetta, M., Shulman, G.L.: Control of goal-directed and stimulus-driven attention in the brain. Nat. Rev. Neurosci. **3**, 201–215 (2002). doi:10.1038/nrn755, http://dx.doi.org/10.1038/nrn755
9. Judd, T., Ehinger, K., Durand, F., Torralba, A.: Learning to predict where humans look. In: 2009 IEEE 12th International Conference on Computer Vision, pp. 2106–2113. IEEE (2009)
10. Gao, D., Mahadevan, V., Vasconcelos, N.: The discriminant center-surround hypothesis for bottom-up saliency. In: NIPS, pp. 497–504 (2008)
11. Le Ngo, A.C., Qiu, G., Underwood, G., Ang, L.M., Seng, K.P.: Visual saliency based on fast nonparametric multidimensional entropy estimation. In: ICASSP, pp. 1305–1308. IEEE (2012)
12. Kadir, T., Brady, M.: Saliency, scale and image description. Int. J. Comput. Vis. **45**, 83–105 (2001). doi:10.1023/A:1012460413855, http://dx.doi.org/10.1023/A:1012460413855
13. Chen, X., Huang, H., Zheng, H., Li, C.: Adaptive bandwidth mean shift object detection.In: RAM, pp. 210–215 (2008)
14. Comaniciu, D., Ramesh, V., Meer, P.: Real-time tracking of non-rigid objects using mean shift. In: IEEE Conference on Computer Vision and Pattern Recognition. Proceedings, vol. 2, pp. 142–149. IEEE (2000)
15. Chen, X., Zhou, Y., Huang, X., Li, C.: Adaptive bandwidth mean shift object tracking. In: 2008 IEEE Conference on Robotics, Automation and Mechatronics, pp. 1011–1017 (2008). doi:10.1109/RAMECH.2008.4681484
16. Vaswani, S., Thota, R., Vydyanathan, N., Kale, A.: Fast 3D structure localization in medical volumes using cuda-enabled gpus. In: 2012 2nd IEEE International Conference on Parallel Distributed and Grid Computing (PDGC), pp. 614–620 (2012)
17. Nvidia cuda c programming guide: http://developer.download.nvidia.com/compute/cuda/32/toolkit/docs/CUDA
18. Hong, W., Georgescu, B., Zhou, X.S., Krishnan, S., Ma, Y., Comaniciu, D.: Database-guided simultaneous multi-slice 3D segmentation for volumetric data. In: Computer Vision-ECCV 2006, pp. 397–409. Springer (2006)
19. Ming-Kuei, Hu: Visual pattern recognition by moment invariants. IRE Trans. Inf. Theory **8**, 179–187 (1962)
20. Chan, C., Kim, J., Feng, D., Cai, W.: Interactive fusion and contrast enhancement for whole body pet/ct data using multi-image pixel composting. In: Nuclear Science Symposium Conference Record, 2005 IEEE, vol. 5, pp. 2618–2621. IEEE (2005)
21. Maalej, N.: Image quality in different modalities of medical imaging with focus on mammography. Physics Department (2004)
22. Jackson, S., Thomas, R.: Introduction to CT Physics
23. Menze, B., Jakab, A., Bauer, S., Reyes, M., Prastawa, M., Van Leemput, K.: Multimodal brain tumor segmentation. MICCAI Challenge. http://www.imm.dtu.dk/projects/BRATS2012

# Missing Data Interpolation Using Compressive Sensing: An Application for Sales Data Gathering

**S. Spoorthy, Sandhyasree Thaskani, Adithya Sood, M. Girish Chandra and P. Balamuralidhar**

**Abstract** Tasks like survey analysis involve collection of large amounts of data from different sources. However, there are several situations where exhaustive data collection could be quite cumbersome or infeasible. In this paper, we propose a novel Compressive Sensing (CS)-based framework to recover the original data from less number of collected data points in the case of market or survey research. We utilize the historical data to establish sparsity of data, and further introduce the concept of logical proximity for better recovery results. Additionally, we also present a conceptual idea toward adaptive sampling using data stream sketching, which suggests whether the collected data measurements are sufficient or not. The proposed CS-based methodology is tested with toy-sized examples and the results are presented to demonstrate its utility.

## 1 Introduction

Analysis of data for a particular trend or feature requires collection of huge volumes of data from different sources. In domains like survey research and market research, data collection plays an important role in further data analysis. Often exhaustive

---

S. Spoorthy · M.G. Chandra (✉) · P. Balamuralidhar
TCS Innovation Labs, Bangalore 560 066, India
e-mail: m.gchandra@tcs.com

S. Spoorthy
e-mail: s.spoorthy@tcs.com

P. Balamuralidhar
e-mail: balamurali.p@tcs.com

S.Thaskani
TRDDC, Pune, India
e-mail: thaskani.sandhya@tcs.com

A. Sood
Indian Institute of Management, Indore, India
e-mail: p13adityas@iimidr.ac.in



R. Singh et al. (eds.), *Machine Intelligence and Signal Processing*,
Advances in Intelligent Systems and Computing 390,
DOI 10.1007/978-81-322-2625-3_3

data collection in these domains is hindered by the inability to collect data from all possible sources, due to reasons like non-availability of manpower or difficulty in reaching out to all sources. As data from all possible sources is necessary for subsequent analysis in such domains, it is necessary to obtain the details of missing data either through estimation or interpolation.

Statistical methods like Maximum Likelihood (ML) estimation and Bayesian Multiple Imputation (MI) are popularly used methods for the estimation of missing data. ML estimation using Expectation Maximization (EM) algorithm has been widely used to estimate unobservable quantities from the collected data (latent data analysis) while imputation methods were alternatives to ML estimates in filling missing data. These methods have performed well in a range of survey analysis tasks [1, 2].

In this paper, we focus on the problem of obtaining details of uncollected data from a set of collected measurements. We present an alternative novel data-independent generic framework based on the Compressive Sensing (CS) paradigm. Compressive Sensing (or Compressed Sensing) (CS) is a sampling theory which recovers the original signal with the help of few non-adaptive linear measurements of the signal. The theory states that a few linear *measurements* obtained from data with a sparse or compressible representation in its native domain or in any transform, using a random sensing matrix, are sufficient to recover the original data using basis pursuit or other greedy algorithms [8].

The CS framework was used in imaging and communication theory applications to collect, retrieve, and transmit huge amounts of discrete data using fewer compressive sensing measurements. It demonstrated reasonable success in areas ranging from hyperspectral image recovery [7] to reliable data transmission [4]. The compressive sensing flavor helped in providing an altogether different perspective, be it for data collection or for data transmission. In this paper, we use the CS paradigm to recover missing data in a survey or market research. For this, we articulate the problem as follows: Let us assume that the original data consists of data from all data sources of interest. The collected or available data is assumed to be a subset of this original data obtained through randomized selection process. By this, we mean that the collected data is treated as compressive sensing measurements obtained from the original data through a randomized sampling procedure. We demonstrate that the original data of interest can be obtained to a good accuracy subjecting the collected data to $l_1$-*minimization* algorithm. By exploiting the linearity of the CS framework, we suggest how to achieve uniform random sampling, which is essential for good reconstruction accuracy. The design of the randomized sampling matrix is the pivot upon which the compressive sensing paradigm is established as data recovery procedure for missing data problems. We also place an additional and crucial premise of *logical proximity* on top of the data sparsity concept of CS theory to improve the sparseness of the data in the transform domain. By this concept, we imply that data sources with logical data patterns can be grouped together to introduce greater sparsity in the sparse domain. In the whole process, we visualize the sales data of a particular product as an image frame (2D matrix) and the data for different products as a group of frames. We also cross-check the sufficiency of the number of measurements by utilizing data stream sketch; if the number of measurements happens to be less, then additional data can be collected in an "adaptive" way. To further aid the analysis, we visualize the data collected from every source at regular periods as a 2-D matrix. This paper, being based on our earlier patent [10], spell out the aforementioned aspects, which to the best of our knowledge are novel.

The rest of the paper is organized as follows: In Sect. 2, a brief description of CS theory is provided. In Sect. 3, compressive sensing framework for survey or market research is described. We further describe the concepts of *exploiting historical data for sparsity* and *logical proximity* in Sects. 4 and 4.1. Later, we focus on sketching methods to find out the sufficiency of measurements required for efficient data recovery in Sect. 5. Some discussion and results are provided in Sects. 6 and 7. The conclusions are provided in the final section.

## 2 Compressive Sensing

Consider a real signal $\boldsymbol{x}$ such that $\boldsymbol{x} \in R^N$ and let us assume it is $k$-sparse (or $k$-compressible) in a known domain $\Psi$, so that it is represented by the following equation:

$$\boldsymbol{x} = \boldsymbol{\Psi s} \tag{1}$$

The domain $\boldsymbol{\Psi}$ can be the signal's native domain or some transform domain. Here, $\boldsymbol{s}$ is the representation of $\boldsymbol{x}$ in the domain $\boldsymbol{\Psi}$. This means that $\boldsymbol{x}$ can be represented using $k$ coefficients (or $k$ largest coefficients, in case of compressible signals) of $\boldsymbol{s}$ in $\boldsymbol{\Psi}$ domain.

According to CS theory, a few linear non-adaptive measurements, $\boldsymbol{y}$, obtained from the native signal $\boldsymbol{x}$ using a random sensing matrix $\boldsymbol{\Phi}$ as

$$\boldsymbol{y} = \boldsymbol{\Phi x} = \boldsymbol{\Phi \Psi s} \tag{2}$$

are sufficient to recover the original signal with near-optimal accuracy. Gaussian random or the Bernoulli random matrices are popularly used random sensing matrices in the CS literature. These random sensing matrices need to satisfy 'Incoherence' property and 'Restricted Isometry Property (RIP)' for successful recovery of the original signal $\boldsymbol{x}$ from the collected measurements $\boldsymbol{y}$. With these random measurements, the factor $\boldsymbol{A} = \boldsymbol{\Phi\Psi}$ also satisfy the above-mentioned properties. When the above conditions are satisfied, the original signal $\boldsymbol{x}$ can then be recovered by subjecting the measurements $\boldsymbol{y}$ to basis pursuit or any other pursuit algorithm, together with the appropriate basis.

## 3 Compressive Sensing for Surveying

We assume that for a certain survey or market research, data needs to be collected from N sources, say, N shops for a given product. This data can be straightforwardly arranged in an $N \times 1$ vector, represented by $\boldsymbol{x}$. Let us now suppose that a small number of these sources have been visited for data collection due to geographical or manpower constraints. We arrange these sources in an $M' \times 1$ vector, represented by $\boldsymbol{y}'$. We can suppose that the measurement vector $\boldsymbol{y}$ is obtained from $\boldsymbol{x}$ due to semi-random sampling. This can be written as

$$\boldsymbol{y}' = \varphi \boldsymbol{x} \tag{3}$$

It is to be noted that we are using different $\varphi$ in Eq. (3) compared to (2) to signify the fact that due to practical limitations and nature of the problem we are considering this matrix that need not be random in the sense of providing uniform random samples. By exploiting the linearity framework of the CS, we can obtain improvised random samples by operating $y'$ with a random matrix of dimension $M \times M'$ to arrive at the "final" measurement vector,

$$\boldsymbol{y} = \boldsymbol{\Theta}\,(\varphi)\,.\boldsymbol{x} = \boldsymbol{\Phi}\boldsymbol{\Psi}\boldsymbol{s} \tag{4}$$

It is important to note that this operation by $\boldsymbol{\Theta}$ on $y'$ is being carried out at the reconstruction end, before feeding it to a pursuit algorithm. This process is really useful since more "effective" randomness in the measurements at the reconstruction side lead to better results. This obviously implies having more measurements to start with; however, keeping with the practicality of data collection mechanism, this still provides a neat way of utilizing the CS framework.

With the mapping to the CS framework being spelt out before applying the CS techniques, we reiterate that it is convenient to visualize and handle the data as image and sequences of images. In this context, we can visualize the data vector $\boldsymbol{x}$ in terms of a data matrix of size $L \times L$, where $N = L^2$. This facilitates arriving at the sparsifying basis with ease and also to work out the "logical proximity," which is elaborated in the next section. Further, when we consider the data for different products, we can treat the data for each product as a frame and collectively consider the "Group of Images" to exploit the correlation among the sales of the products, to finally reduce the amount of data collected. Operating the image matrix with the CS model can also be carried out as suggested in [3, 4]. The reader is referred to Sect. 6 for additional details. Headings should be capitalized (i.e., nouns, verbs, and all other words except articles, prepositions, and conjunctions should be set with an initial capital) and should, with the exception of the title, be aligned to the left. Words joined by a hyphen are subject to a special rule. If the first word can stand alone, the second word should be capitalized.

## 4 Sparsity Basis and Logical Proximity

Once the measurement (data) is collected, we can randomize it by $\boldsymbol{\Theta}$ and feed it to CS reconstruction algorithms. These algorithms also require the basis in which the data is sparse to carry out the reconstruction. Since we have data model as images, we can try Discrete Cosine Transform (DCT) or Discrete Wavelet Transform (DWT) matrices. Of course, if some historical or similar data is available, one can also establish the appropriate sparsifying basis, by analyzing the data as well.

### *4.1 Logical Proximity for Efficient Data Recovery*

With the historical data available, one can try grouping the data with similar sales patterns as nearby pixels in the image frame, rather than lexicographical arrangement. This results in better intra-frame correlation, and hence better sparsity. This can be either exploited to give better reconstruction accuracy for a given number of measurements or reduced number of measurements for a given accuracy. A combination trade off is also possible.

## 5 Thoughts on Obtaining Requisite Measurements Using Data Stream Sketching

With the historical data suggesting a guidance on logical proximity and sparsity, we can arrive at the number of measurements using the CS theory. But it is to be noted that the sales data gathering scenario is dynamic, which can disturb the sparsity over

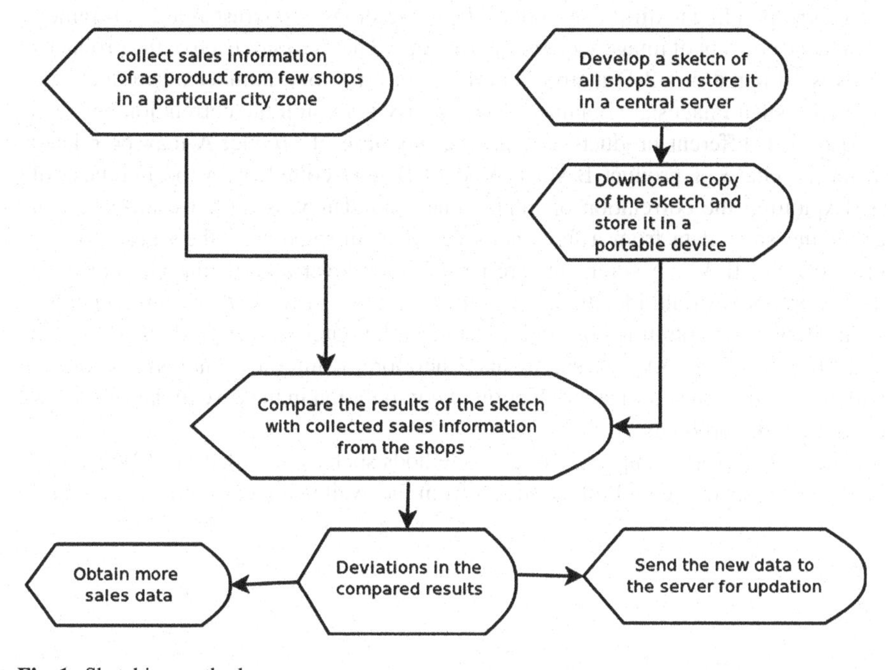


**Fig. 1** Sketching method

time. One cannot be sure that the number of measurements sufficient previously is enough now also, or if it is required to collect more data in the changed circumstance. In order to arrive at the latter decision, it is required to compare the CS reconstructed results with the "correct" results. In the large-scale scenario, which is typical of sales data gathering, let us hypothesize that we can create a summary of the entire sales data and have access to that summary alone. One way to create this summary is through data stream sketching. The reader is referred to [5, 6]. The analyst or firm interested in validating the data gathering through CS can download the sketch and compare the requisite number of query results with CS reconstruction results as depicted in Fig. 1. If there is considerable discrepancy, one can collect more data, leading to adaptive sampling within our scenario. Furthermore, sketches and the associated results can also guide us in reworking the logical proximity in an "evolutionary" fashion.

# 6 Discussion

## *6.1 Extension of the Case to Multiple Products*

In Sects. 3 and 4, we proposed the CS model for data recovery for a single product for a large set of shops, given data from few shops. In this section, we extend the logic to sales data recovery in the case that data corresponds to the sales of multiple products. In such a case, as was mentioned earlier, we propose to visualize the data corresponding to different products from the shops of interest as a set of frames, akin to the concept of image sequences in a video. That is to say, each of the product's sales is arranged in a 2-D matrix, and therefore, if a total number of products are $S$, then $S$ such images are obtained. Now, we try to exploit the correlation between the sales of different products as thus, the purchase of product A may be related to the purchase of product B since product B may complement the purchase of A. Exploiting the correlation or inter-frame redundancy as thus, we observe that collecting sales data from a few shops can help in relating to the sales of other products too. If $M$ measurements are needed to recover a particular class of data, then $S$ products would ideally result in the collection of $S \times M$ measurements per shop. However, exploiting correlation between the purchases of products results in collection of $\tilde{M} \ll SM$ measurements. Therefore, using the compressive sensing paradigm results in significant reduction of computation in the case of data recovery using multiple products.

Thereafter, video compression-based methods such as provided by [12] can then be used to recover the multiple products from the available measurements or data.

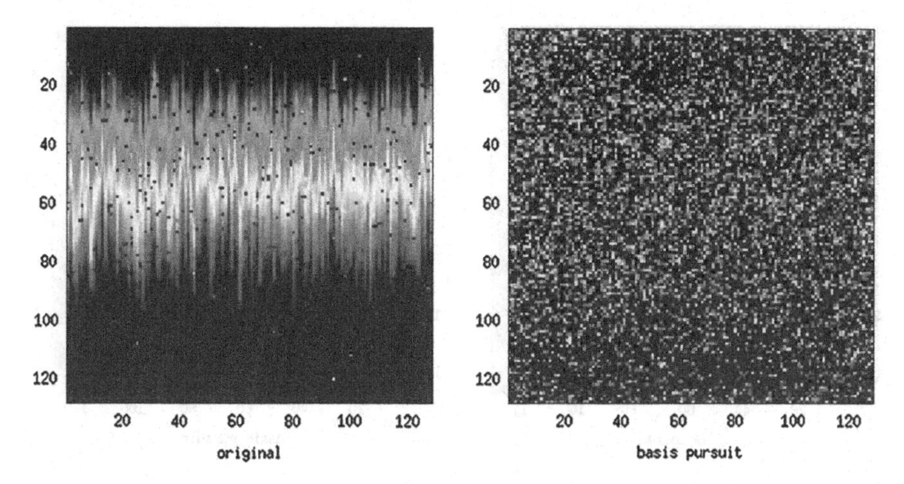


**Fig. 2** Recovery of sales data in the case where logical proximity is not considered

# 7 Results

We substantiate the working of the proposed model in Sect. 3, with the help of toy-sized simulation results for the sales data recovery. We consider data collection from $M$ shops, where the sampling matrix $\varphi$ is semi-random. To simulate this scenario, we forced an intentional choice of a few shops say $P$, where $P < M$, and selected rest of the shops in a random manner. Due to the forced choice of few shops, $\varphi$ had to be subjected to two stages of randomization in order to achieve the final measurement matrix, described in (2). Therefore, we find that the randomization operator $\Theta$ given by (4) is the product of a two-stage randomization process here. The sparse domain for this data was DCT domain. In this paper, we present the results for sales data recovery in the case of a single product. We present the results of two cases: one without logical proximity consideration and the other considering logical proximity.

In the first case, similar sales patterns are not considered for the arrangement of sales data. Figure 2 provides the results of this case. The simulation provided in Fig. 3 is for the case where logical proximity or sales data pattern is considered for data recovery. We find that logical proximity induces a higher degree of sparsity into the data, and therefore provides for near-optimal recovery.

In both illustrations, DCT domain was the sparse domain. For the purpose of illustration, we considered an original data matrix which is a set of 16384 shops, arranged in a $128 \times 128$ matrix. We assumed that only 5504 shops have been visited for data collection. Therefore, the measurement matrix is a $43 \times 128$ matrix.

Based on the proposition in Sect. 3 and simulations performed to illustrate this proposition (Figs. 2 and 3), we observe that data recovery, in most practical situations, may not be optimal owing to two reasons:

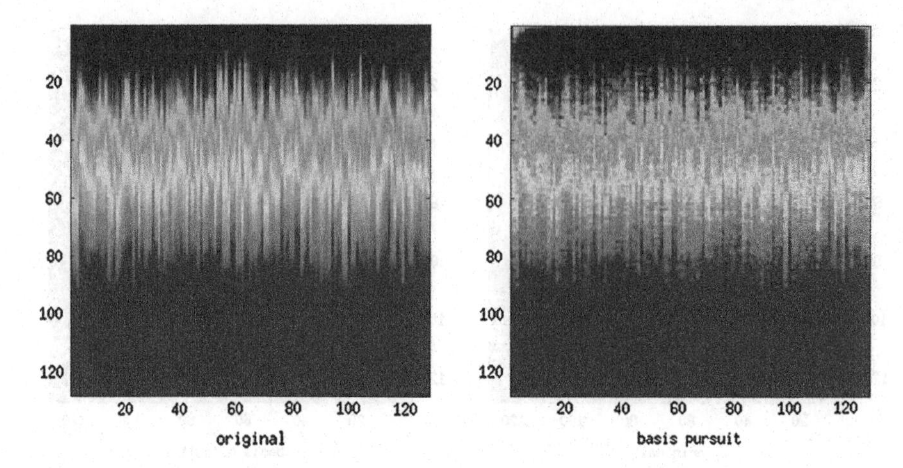


**Fig. 3** Recovery of sales data in the case where logical proximity is considered

1. Firstly, it may not possible to establish strict sparsity for the data under consideration, that is, the insignificant transform-domain coefficients may have small amplitudes, but are not zero.
2. Secondly, in the case of semi-random sampling given by Eq. (3), small amount of uniformity of selection may persist in the final measurement matrix, after the process given by (4) is carried out. This factor may also contribute to sub-optimal solutions.

# 8 Conclusions

In this paper, we presented a novel framework of obtaining missing data from a limited set of measurements using compressive sensing paradigm, in the case of single and multiple products for survey or market research. Towards this end, we present a novel methodology to obtain or collect samples using semi-random sampling matrix and to further improve the "effectiveness" of this matrix at the receiver's end with the help of a randomization operator. We further demonstrated the use of *logical proximity* concept, with the help of toy-sized examples, in improving the efficiency of data recovery. Moreover, we presented an adaptive sampling scheme to check the sufficiency of collected measurements during data updations, using the data stream sketching method. Thus, this methodology along with the proposed CS technique ensures smooth data collection for near-optimal recovery of data despite changes or updates in the data. We have demonstrated the working of the framework and the included methodology with the help of toy-sized examples for single-product data collection case. In summary, the proposed CS methodology is sure to provide an excellent framework to deal with large data recovery problems in survey or market research, due to the flexibility it provides at data collection and updation stages.

## References

1. Rubin, D.B.: Inference and Missing Data. Biometrika **63**(3), 581–592 (1976). doi:10.1093/biomet/63.3.581. Oxford University Press
2. Schafer, J.L., Graham, J.W.: Missing data: our view of the state of the art. Psychol. Methods **7**(2), 147–177 (2002). http://dx.doi.org/10.1037/1082-989X.7.2.147
3. Pudlewski, S., Melodia, T.: On the performance of compressive video streaming for wireless multimedia sensor networks. In: Proceedings of IEEE International Conference on Communications (ICC), Cape Town, South Africa, May (2010)
4. Kadhe, S., Thaskani, S., Chandra, M.G., Adiga, B.S.: Reliable data transmission in sensor networks using compressive sensing and real expander codes. In: 2012 National Conference on Communications (NCC), pp. 1–5, 3–5 February 2012. doi:10.1109/NCC.2012.617684
5. Aggarwal, C.C., Yu, P.S.: A survey of synopsis construction in data streams. Data Streams. Advances in Database Systems, vol. 31, pp. 169–207. Springer, New York (2007)
6. Adiga, B.S., Shastry, R., Chandra, M.G., Rajan, M.A.: A reversible sketch based on chinese remainder theorem: scheme and performance study. IJCSNS **11**(8), 59 (2011)
7. Golbabaee, M., Arberet, S., Vandergheynst, P.: Multichannel compressed sensing via source separation for hyperspectral images. In: Proceedings of Eusipco 2010. No. EPFL-CONF-149116 (2010)
8. Candes, E.J., Wakin, M.B.: An introduction to compressive sampling. IEEE Signal Process. Mag. **25**(2), 21–30 (2008)
9. Baraniuk, R.G.: Compressive sensing. IEEE Signal Process. Mag. **24**(4) (2007)
10. Thaskani, S.S., Sood, A., Purushothaman, B., Chandra, M.G.: A method and a system to conduct a survey. Application no: 20140149181, Filed: Nov 26, 2013, Issued: May 29, 2014: Assignee: Tata Consultancy Services Limited (Mumbai). Application Serial: 14/090,583
11. Bin, G., Yang, Z.: Cooperative compressed sensing for wide-band spectrum detection with sequential measurements. In: Proceedings of the 12th IEEE International Conference on Communication Technology (ICCT) (2010)
12. Secker, Andrew, Taubman, David: Lifting-based invertible motion adaptive transform (LIMAT) framework for highly scalable video compression. IEEE Trans. Image Process. **12**(12), 1530–1542 (2003)
13. Candès, E.J.: Compressive sampling. In: Proceedings of the International Congress of Mathematicians: Madrid, 22–30 August 2006: invited lectures (2006)
14. Adiga, B.S., Chandra M.G., Shastry R., Kadhe, S.: Data Streams and Sketching. TCS Technical Report (July 2011)

# Domain Adaptation for Face Detection

**Vidit Jain**

**Abstract** Face detection has been an active area of research for several decades. As a result, efficient algorithms and implementations have been developed for several practical applications. These advancements have led to significant increase in the scope of face detection research. Adapting existing detectors to different yet related domains is one such extension over the traditional scope. In this setup, the adaptation is performed when only one or a few images are available from the target domain, whereas none of the images from the source domain are available. This chapter discusses two of the recent algorithms that address this problem.

**Keywords** Face detection · Domain adaptation

## 1 Introduction

*Face detection* refers to the problem of determining the location and extent of *all* of the faces present in an image. Also, no assumption is made about the number of faces present in the given image. In single-image face detection, this image is analyzed in isolation, and is *not* assumed to be a part of a sequence (or collection) of images or a video. The discussion below focuses on single-image face detection. Hence, it ignores solutions that exploit the continuity in the location, size, and appearance of face regions in consecutive images in a sequence. In the absence of such structured information, detecting faces in single image is likely to be more difficult than detecting faces in sequence of images.

Detecting faces in single image is useful in a variety of real-world applications. For instance, face detection systems are used for automatically controlling the focus and exposure in digital cameras [1], and for categorizing and filtering retrieved images by commercial image search engines such as Google and Bing. These widespread applications have led to significant developments in the different aspects of performance

V. Jain (✉)
IIIT-Delhi, New Delhi, India
e-mail: viditumass@gmail.com



R. Singh et al. (eds.), *Machine Intelligence and Signal Processing*,
Advances in Intelligent Systems and Computing 390,
DOI 10.1007/978-81-322-2625-3_4

**Fig. 1** *Typical failure cases*. Most of the automated face detectors work well only on faces of persons from similar race and age groups as those included in the training examples. The detected faces are outlined by *green rectangles* in these images

of the face detection systems. Some algorithms focused on increasing the speed of detection [2], whereas others reduced the time taken for training the detector [3]. There has also been some work on increasing the range of detectable variations in face appearance (e.g., across poses [4]). The resulting face detectors perform extremely well within its pre-defined scope of acceptable appearances. However, they fail miserably in unconstrained environments (e.g., see Fig. 1). The low performance of most of these face detectors is perhaps due to the fact that they perform the face versus non-face classification independently for each candidate image region. Learning a precise model of this distinction between face and non-face regions requires a large number of accurately labeled examples of face regions to capture the diverse variations in appearances with different expressions, poses, occlusions, and illumination patterns. The huge cost (both time and effort) incurred in obtaining these labeled examples becomes the bottleneck in expanding the scope of the learned models. On one hand, it is infeasible to collect labeled examples for every domain of face appearances; hence, effective detectors cannot be trained separately for each domain. On the other hand, the trained detectors fail to perform well even on face regions whose appearance is slightly different from those in the domain spanned by the training instances.

Humans are extremely good at detecting human faces in an image. Studies suggest that human brain uses context to solve difficult cases of detection (see Jain's thesis for a detailed discussion [5]). However, the automated systems typically focus only on a given image region, ignoring any kind of information external to the appearance of this image region. As a result, these systems are trying to solve face detection problems without using any scene context. In other words, they are solving a problem that may be more difficult than necessary. This chapter discusses two recent approaches that are inspired by this observation and employ the context about the image or the different image regions to improve the performance of a pre-trained detector in a new domain. Section 3 presents an approach that considers every image as a new domain and performs adaptation on-the-fly without any labeled examples. Section 4 presents a one-time adaptation of a pre-trained detector to build a detector for a new, related domain for which only a few labeled examples are available. The next section describes a common approach for implementing face detection algorithms and provides the necessary background for the above two approaches.

## 2 A Common Approach

One common approach for implementing face detection algorithm follows three steps: (illustrated in Fig. 2):

1. Sample candidate regions from the given image,
2. Classify each of these image regions as a face or non-face region, and
3. Post-process the detected face regions to merge overlapping detections and remove spurious detections.

Modeling the dependencies among these three steps is avoided mostly due to the high computational cost associated with them. The discussion below focuses on developing better alternatives for the second step i.e., face classification. For other two steps, the reader is advised to follow standard techniques [6].

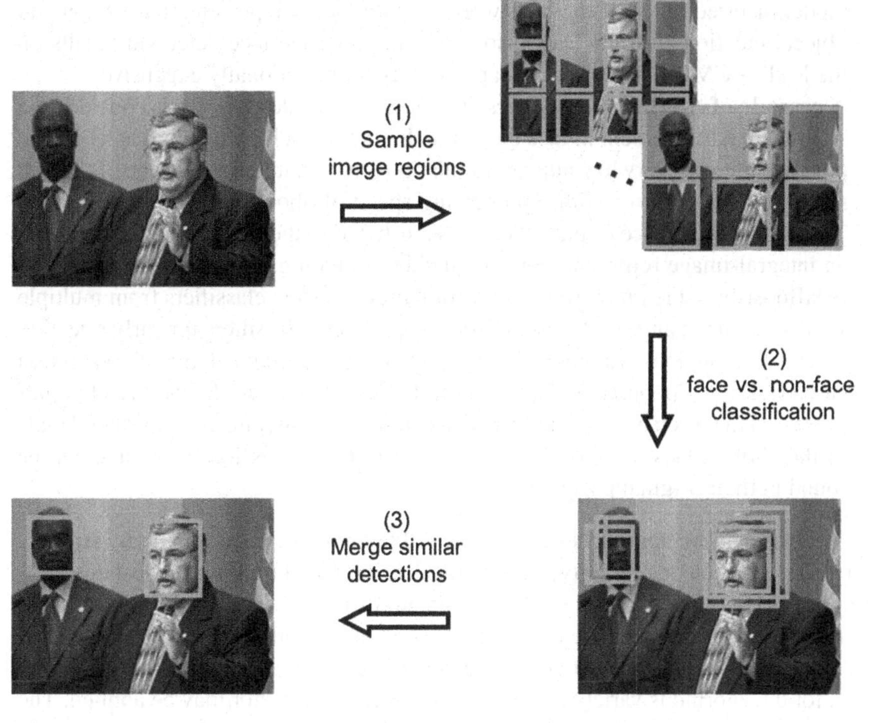


**Fig. 2** *A common implementation of a face detector*

## 2.1 Face Versus Non-face Classification

A variety of statistical techniques have been explored for the binary classification task of determining whether a given image region is a face region. A few face detectors that represent key progress in this area of research are summarized below. The reader is referred to Zhang and Zhang [7] for detailed survey of face detection research. In particular, three approaches are discussed that achieved very impressive results, and provided useful baselines for evaluating other face detection algorithms.

- **Neural networks**. Rowley et al. [8] trained several multi-layer perceptrons with different receptive fields to predict the classification label for a given image region. An arbitration perceptron is trained to combine the output of all of these trained neural networks for generating the final prediction. While they achieved impressive results on the images from the MIT + CMU dataset, it remains unclear how to determine a good set of parameters (e.g., number of layers and hidden units) for their neural network to perform well on a new image collection.
- **Parts-based model**. Schneiderman and Kanade [9] formulated the problem of face detection as the determination of the maximum a posteriori estimate of the presence of a face given the observed image patch. In their model, an image region is represented as an ordered set of multiple subregions. Intuitively, these subregions correspond to different parts of a face such as the eyes and the nose. This model inspired much of the later work on parts-based representation for general object detection as well. They also showed impressive face detection results on the MIT + CMU dataset, but this approach is computationally expensive.
- **A cascade of AdaBoost classifiers**. The Viola–Jones detector [6] is considered a significant advancement in face detection. It not only achieved a high level of true positive rate with very low number of false positives, but it was also one of the first real-time face detector with a processing speed of about 15 frames per second. The high performance of this detector is attributed to the following: (1) the use of an integral-image representation for quickly evaluating feature responses; (2) an AdaBoost-based learning framework for building strong classifiers from multiple weak classifiers; and (3) the use of a cascade of these classifiers for early rejection of non-face regions. Figure 3 shows a schematic diagram of this classification cascade. Being a popular implementation choice, the domain adaptation techniques presented in this chapter are described in detail for this architecture. Further details of the Viola–Jones detector are not relevant to the discussion below and can be found in their original paper [6].

All of the above approaches are supervised learning techniques and rely strongly on the assumption of similarity between the distribution of training and test domains. In practical settings, there are often significant differences between these distributions. These differences arise due to the cost (or infeasibility) of collecting labeled examples that include the test domain, that is, it may be infeasible to collect training data for the enormous variety of domains in which the detector may be applied. The next section considers each test image as a new domain and presents an algorithm

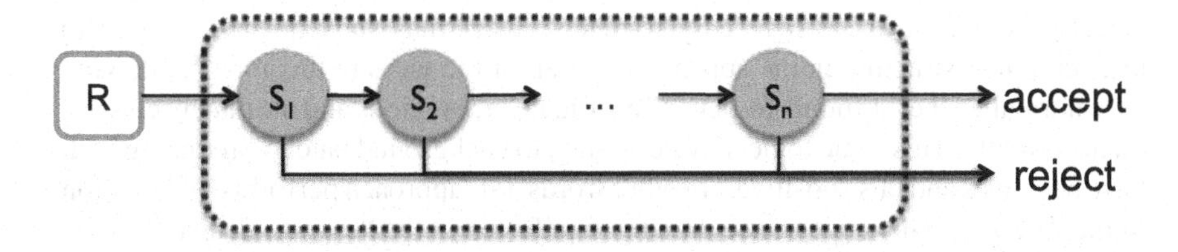


**Fig. 3** *The classification cascade of the Viola–Jones face detector*. Given an image patch $R$, the Viola–Jones face detector uses $n$ binary classifiers to define a cascade that instantaneously rejects a patch that is rejected by any of the $n$ classifier. As a result, an image region is classified as a face region if and only if it is accepted by all the classifiers in the cascade

to adapt a pre-trained classifier to improve its performance on the given image. The underlying principle could be applied to any choice of face classifier, although the details are presented for adapting the classification cascade of the Viola–Jones detector.

## 3 Single-Image Adaptation

The faces appearing in a single image have a lot of similarity in their appearance because they often share a common geometric structure, illumination sources, and other properties of the scene. For instance, all of the face regions in Fig. 4 have similar shadow patterns. Due to the difference in the pose and position of these faces relative to the camera, some faces have weaker shadows compared to the other faces. An automated detector is likely to fail on the face of the person sitting in the left-bottom corner because of the shadow on the left half of his face. Since the shadow is relatively small on two faces in the right half of the image, a good detector may successfully

**Fig. 4** *Easy-to-detect faces could help identify hard-to-detect faces*. There is a shadow in the left half of all the four faces. The shadow is stronger on the two faces on the left, making them more difficult to detect than the other two faces. A detector that could learn the shadow pattern from the easy-to-detect faces could normalize the other faces to reduce their difficulty of detection

detect these faces. These two "easy-to-detect" faces may subsequently be used to infer common structure in the appearance of all of the faces in this image, allowing us to normalize the "difficult-to-detect" candidate face regions and ultimately classify them correctly. This same logic may be applied to background patches, reducing both false negatives and false positives. In other words, this approach performs a adaptation of the detector to the appearance domain manifested in the "easy-to-detect" faces.

In this domain adaptation setting, no labeled data is available for the target domain (i.e., a single image). Also, the data used for training the original classifier (i.e., from the source domain) is also unavailable during adaptation. These two constraints make most of the work on domain adaptation [10–14] not relevant in this setting. These approaches assume that a small number of labeled examples are available from the target domain in addition to the labeled data from the source domain. A discussion of these approaches is deferred to Sect. 5.1.

One naïve way to implement the above intuition is to scan the image for high-confidence faces, and then adapt the detection model according to the high-confidence face and non-face regions. This two-step process has two problems. First, it can lead to overfitting the first step's predictions. Second, it represents a substantial increase in computation. Both of these problems can be addressed by formulating face detection as a regression problem and using an efficient Bayesian model with a strong prior.

**A regression-based formulation**. The appearance quality of a face region in an image depends on several factors including the pose of the person, the distance of the face from the camera, and the occlusion of the face from other objects present in the scene. For instance, in Fig. 5, the resolution of the faces of people in the audience is much lower than the resolution of the faces of the players. It is likely that there is little commonality in the structure of appearance between these two classes of faces in this image. In a different image, there may be more of such modes in the distribution of face appearances present in a single scene. The presence of these unknown numbers of different modes of appearance and the potential (gradual) transition among them

**Fig. 5** *Multiple modes in appearance quality of face regions*. The resolution of the faces of people in the audience is much lower than the resolution of the faces of the players. Thus, it is likely that there is little commonality in the structure of appearance between these two classes of faces in this image

makes it challenging to learn the desired adaptation of a face detector in the traditional classification setting. Moreover, the number of confident face and non-face regions in an image may be too small to learn an effective nearest-neighbor based classifier.

A regression-based formulation is useful in this setting. In this formulation, image regions with similar appearance receive similar detection scores (after adaptation). In other words, the detection scores for the faces in the audience are encouraged to be similar to each other but may be different from the detection scores for the face regions corresponding to the players. A similar effect can be achieved using an ensemble of classification models, although specifying a strong prior—to avoid overfitting to the new observations from the "easy-to-detect" faces—becomes computationally prohibitive for them. In the regression setting, a Gaussian process prior [15] provides a straightforward way to implement the corresponding Bayesian model. The mathematical overview of this adaptation is detailed below.

## 3.1 Online Domain-Adaptation

Let $\mathbf{x} \in \mathbf{X}$ denote the feature representation of a data instance (i.e., an image region, in this discussion) and $S$ denote a classifier based on the sign of the prediction value from a function $f(\cdot)$, i.e.,

$$S(\mathbf{x}|f) \triangleq \text{sgn}(f(\mathbf{x})). \tag{1}$$

Assume the probability of error of $f$ to be monotonically non-increasing with the absolute value of $f(\mathbf{x})$, i.e., $|f(\mathbf{x})|$. In other words, the large prediction values (both positive and negative) are more likely to be correct than the small prediction values. This assumption also suggests that the classification label obtained by the classifier $S$ for the points near the classification boundary (i.e., 0) may not be reliable. For the face detection problem, this assumption suggests that the pre-trained classifier can confidently accept unoccluded, in-focus, or "easy-to-detect" faces, and reject several non-face regions from a given image. This classifier is assumed to generate high prediction values for these easy acceptances and rejections. Consequently, the decision for the faces with low prediction values is assumed to be more difficult compared to the decision for the regions with high prediction values.

For the data instances with low prediction values from the pre-trained classifier, new prediction values are computed that encourage consistency among all of the prediction values. In other words, if two data points $\mathbf{x}_1, \mathbf{x}_2 \in \mathbf{X}$ are similar to each other, then the corresponding predictions $f'(\mathbf{x}_1)$ and $f'(\mathbf{x}_1)$ should also be similar to each other. To this end, a small margin around the classification boundary is defined. The data points with prediction values outside this margin are used to learn a Gaussian process regression (GPR) model, which is used to update the prediction values of the data points with prediction values lying inside the margin. Figure 6 illustrates the intuition for this classifier adaptation, and the formal description is included below.

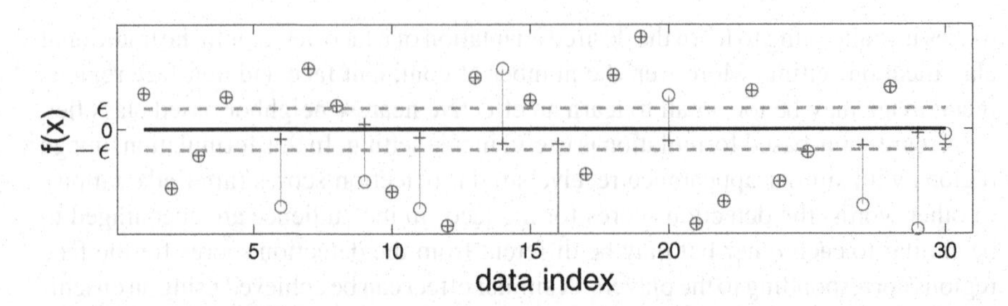


**Fig. 6** *An illustration of online domain adaptation.* $f(x)$ denotes the output of a score predictor on a data point $x$. Consider an $\epsilon$ margin (*green dotted line*) around the classification boundary (*black solid line*). For points lying in the margin, the classifier is not very certain about the predictive label. The proposed method updates the scores for the points in this margin based on their similarity to the other points for which the classifier is relatively more confident about the classification label. The original classification output is shown using *blue* '+,' whereas the updated output (obtained using Eq. 4) is shown using *red* 'o'

Given $\epsilon > 0$, the *in-margin set* $\mathbf{X}_m \subseteq \mathbf{X}$ is defined as

$$\mathbf{X}_m \triangleq \{\mathbf{x} \in \mathbf{X} \text{ s.t. } |f(\mathbf{x})| < \epsilon\}. \tag{2}$$

Similarly, the *out-of-margin set* $\mathbf{X}_o \subseteq \mathbf{X}$ is defined as

$$\mathbf{X}_o \triangleq \mathbf{X} \setminus \mathbf{X}_m. \tag{3}$$

Jain et al. [16] learned a Gaussian process regression model using the instances in the out-of-margin set $\mathbf{X}_o$. The mean and variance functions of the learned model are used to compute the update function $f'(\cdot)$ as

$$f'(\mathbf{x}) = \begin{cases} f(\mathbf{x}) & \text{if } \mathbf{x} \in \mathbf{X}_o \\ \mu(\mathbf{x}, \mathbf{X}_o, \Phi(\mathbf{X}_o)) - c\sigma(\mathbf{x}, \mathbf{X}_o) & \text{otherwise.} \end{cases} \tag{4}$$

The final updated classifier is defined as

$$S'(\mathbf{x}|f, \epsilon) \triangleq \text{sgn}(f'(x)). \tag{5}$$

## 3.2 Cascade of Adapted Classifiers

As discussed in Sect. 2.1, Viola–Jones cascade of classifiers (Fig. 3) is an effective architecture for implementing a face detector. Algorithm 1 describes the update mechanism for such a cascade of classifiers using the above method of adapting a pre-trained classifier. For this algorithm, the input $X$ refers to the set of candidate image regions to be evaluated by the cascade. The margin parameter $\epsilon$ is chosen by

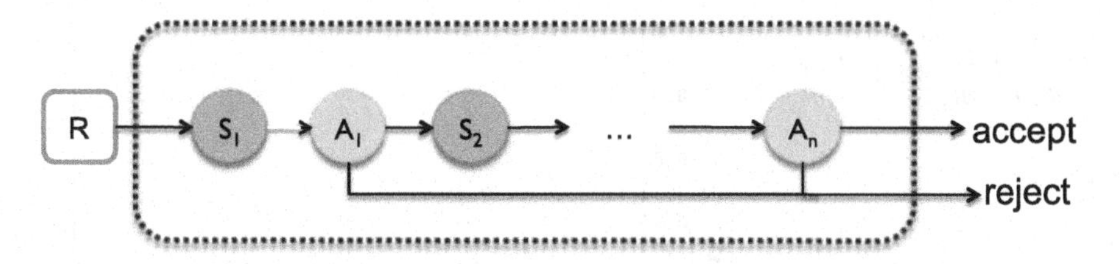


**Fig. 7** *The adapted cascade.* The stage classifiers from the original cascade are denoted by the *blue circles*, whereas the score update functions are denoted by the *green circles*. Each stage in the adapted cascade corresponds to a pair of a *blue* and a *green circle*. At each stage, the scores from the original stage classifier are used to learn a score update function. The updated scores are then used to make the decision to reject the candidate image regions

maximizing the true positive rate at the observed number of false detections equal to 10 % of the total number of true faces on a held-out dataset.[1] The resulting cascade is referred to as a *cascade of adapted classifiers* (Fig. 7).

**Algorithm 1** Cascade of adapted classifiers.

**Require:** input $X$, classifier cascade $\{S\}_{1\ldots n}$, margin $\epsilon \geq 0$

1: **for** $n = 1$ to $N$ **do**
2: Let the stage classifier $S_n := \text{sgn}(f_n(x))$
3: $X_m \leftarrow \{x \in X \mid |f_n(x)| < \epsilon\}$
4: $X_o \leftarrow X \setminus X_m$
5: $Y_o \leftarrow f_n(X_o)$
6: train a GPR model using $X_o, Y_o$
7: $\forall x \in X_o, f_n'(x) \leftarrow f_n(x)$
8: $\forall x \in X_m$, compute $f_n'(x)$ using Eq. 4.
9: $X \leftarrow \{x \in X \mid f_n'(x) > 0\}$
10: **end for**

## *3.3 Observations*

Despite the maturity of face detection research, it has been challenging to compare different algorithms for face detection. The traditional datasets (e.g., the MIT+CMU dataset) provided good sets of images for training and testing but did not provide a well-defined strict protocol. Also over time, several competing algorithms achieved extremely high performance on this dataset. To address these problems, Jain and Learned-Miller develop the face detection dataset and benchmark, commonly known as FDDB [17]. This dataset included challenging images from real-world

[1]This strategy is similar to the suggestion made by Viola and Jones [6] for selecting certain model parameters.

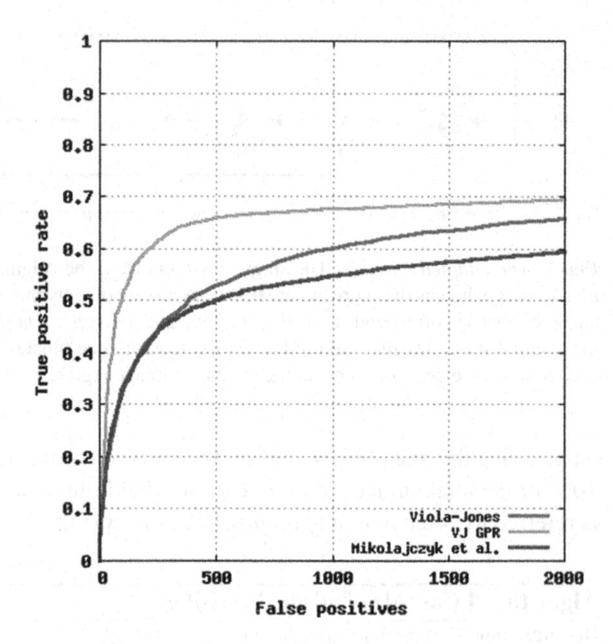


**Fig. 8** *A comparison of different face detectors based on the FDDB discrete score*

scenes, set a strict protocol for reporting performance, and also maintained a webpage of performance curves of different algorithms. Jain and Learned-Miller [16] reported their observation of single-image adaptation using the EXP-2 experimental setup of the FDDB benchmark. They used the OpenCV[2] version 1.0 implementation of the Viola–Jones frontal-face detector as the original, pre-trained cascade. Figure 8 presents the performance curves for Viola–Jones detector, Mikolajczyk et al.'s face detector [18] and the above adapted cascade-based face detector. The above-mentioned adaptation is performed on the Viola–Jones detector and Gaussian process regression is used for re-computation of detection scores; therefore, the adapted cascade-based detector is referred to as *VJ GPR*. Figure 9 shows the comparison of face detection results obtained by the Viola–Jones detector and the adapted detector with parameter settings such that they obtain same false positive rates on FDDB.

No post-processing of the detected regions was performed in these experiments to improve the absolute performance of these detectors. The focus was on the fundamental improvements in face detection approaches, and similar optimization for multiple poses was also ignored. These results show that a simple, online adaptation of a black-box classifier substantially improves its performance on a new target domain. It would be interesting to evaluate semi-supervised methods in this scenario, but they require the access to the original training data for the Viola–Jones detector. This constraint makes the above approach even more appealing for online domain adaptation.

[2]https://www.sourceforge.net/projects/opencvlibrary/.

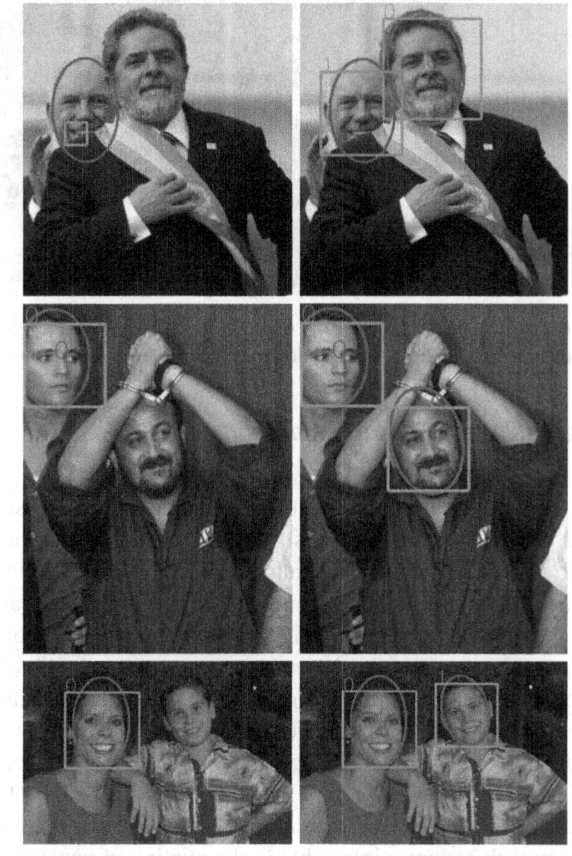


**Fig. 9** *Detections obtained by two face detectors on some images from FDDB*.The detections are denoted by *green rectangles*, whereas the matched ground-truth face annotations are denoted by *red ellipses*. These face detection results are obtained using systems with identical false positive rate

# 4 Detection in New Domains

The previous section described an approach for performing the adaptation using a single image and with no labeled examples. There are several other practical settings where this assumption of extreme lack of resources is not necessary. A few hundred images may be available for domain adaptation and some human annotation effort may be affordable. This section considers this setting of limited availability of training data in a new domain. This setting is relevant not only for face detection but also for several other rare-event classification problems such as intrusion detection. Similar to the earlier discussion, this section presents the adaptation for the pre-trained Viola–Jones cascade.

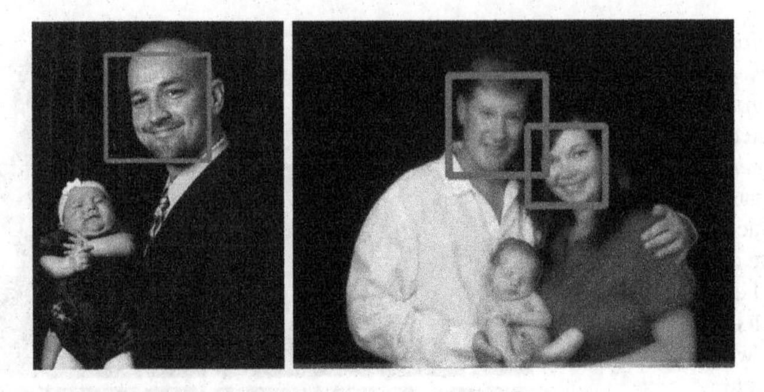

**Fig. 10** *Narrow range of detections*. In both of these images, a standard face detector correctly identified the adult faces but failed to detect the faces of the babies

## 4.1 Cascade Adaptation

Consider the problem of domain adaptation for a cascade of classifiers when the available positive examples from the target class are not sufficient to train cascade from scratch. Also, it is assumed that only the pre-trained cascade is available, and *not* the data used for training it.[3] This pre-trained cascade fails to perform well on images with variations in appearances that may not be captured in the training examples, even when the variations are minor (in their semantics). For instance, a cascade trained on the images of adult-human faces fails to detect faces of babies (see Fig. 10). This limited generalization of a cascade classifier severely restricts the data domains for which it can be used effectively. A common approach to address this limitation is to train a new cascade of classifiers from scratch for each of the new domains. Due to the requirement of a large collection of labeled examples, this approach becomes infeasible as the number of new target domains increases.

This lack of robustness to variations in appearances can be attributed to an overfitting done by the different classifiers in the cascade. A deeper analysis of this overfitting requires a functional understanding of the classifiers in the different parts of this cascade. Formally, a cascade $\mathcal{F}$ is a classifier that is composed of $m$ stage classifiers $\{S_1, \ldots, S_m\}$ that are applied in a sequential manner. As detailed below, the first stages $\{S_1, \ldots, S_h\}$ (*head*) perform an easy rejection of negative examples, whereas the last stages $\{S_{t+1}, \ldots, S_m\}$ (*tail*) perform a more holistic verification of positive examples; the middle stages (*torso*) perform the classification of different structural components of the face at a different scale and scope. This phase-wise split is illustrated in Fig. 11. Each of these three phases requires a different method of adaptation.

**Head adaptation**. These early stage classifiers are expected to eliminate only the easy-to-reject instances, and therefore they can be trained effectively from scratch even with a few training examples from the new domain. The acceptable hit rate

[3]While it is common to make a pre-trained classification cascade available, it is often not feasible to retain the examples used for training it due to operational and copyright issues.

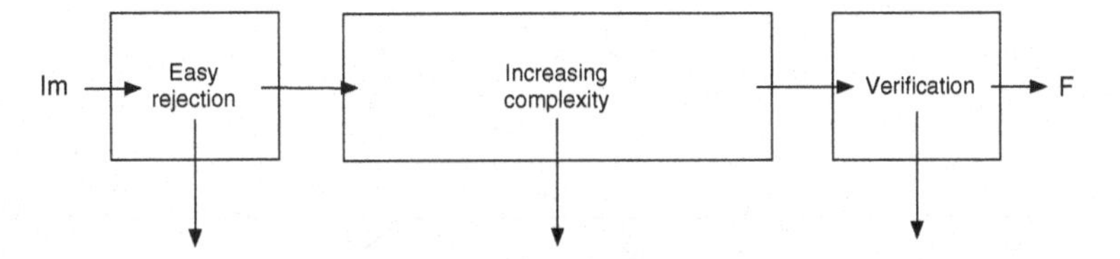


**Fig. 11** Phase-wise split of a cascade of classifiers

and false alarm rate are set similar to the training objective for the original Viola–Jones cascade. The computational efficiency of the original cascade is maintained by considering the same family of classification functions to learn stage classifiers for the new cascade. The stage classifiers from this new cascade replace the $\{S_1, \ldots, S_h\}$ stage classifiers.

**Torso adaptation**. The torso of the cascade is composed of an ordered set of stage classifiers $\{S_{h+1}, \ldots, S_t\}$ of increasing complexity. If any of these stage classifiers reject a given image patch, the patch is immediately discarded, otherwise it is evaluated by the next stage classifier. The complexity of the subsequent classifiers increases because the acceptable false alarm and hit rate for the trained classifier become stricter for subsequent stages. This phase captures different structures of medium complexity in the appearance of the source domain. If the target domain is similar to the source domain, some of these structures are expected to be shared across the two domains. For instance, the presence of a pair of eyes would be shared in faces of adult humans and babies. By selecting only the stage classifiers that capture these shared structures, a new classification cascade is constructed for the target domain. The stages that capture structures which differ between the two domains are removed. For example, the relative ratio of the distance between two eyes to the distance between eyes and lips is expected to be different between adult humans and babies. The stage classifier that examines this ratio should be ignored while adapting the cascade to detect faces of babies. Jain and Farfade [19] implemented this stage selection as follows. For each stage, in this cascade, they introduced a binary selection variable $\theta$ that specifies whether the evaluation of this stage is useful for the target domain. They learned the instantiation of these binary variables using the few labeled examples from the target domain. This model of stage selection is illustrated in Fig. 12. The reader is referred to Jain and Farfade [19] for the details of the parameter estimation for this model.

**Tail adaptation**. The tail stage classifiers, i.e., $\{S_{t+1}, \ldots, S_m\}$, of the cascade are performing a validation of the hypothesized detection through a matching of the detailed structures present in the instances of the given object class. This functionality can alternatively be obtained using a generative modeling of the appearances of the given class instances. This generative model should learn effective parameters from only a few examples and should require similar (or less) computational effort as the tail of the original cascade. Jain and Farfade [19] proposed the use of a nearest-neighbor-based probabilistic model in a projection space for this purpose. In particular, they computed the projection space using the non-negative matrix factorization [20] of

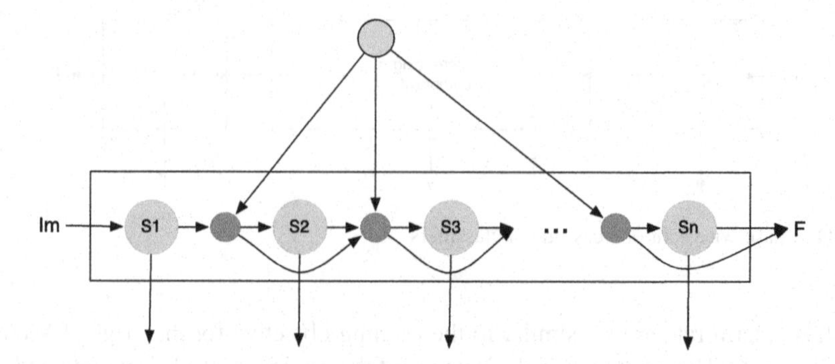


**Fig. 12** *Classification cascade with stage selection*. Each of the stage classifiers (*blue*) has a binary selection variable (*red*) associated with it. These selection variables share the relaxation parameters (*green*)

the training images $X$ from the target domain. In this learned subspace, they employ a non-parametric density estimator using $k$ nearest neighbors for each of the training examples to compute the validation probability of the hypothesized detection.

## 4.2 Observations

One relevant application of this adaptation setting is manifested in image search on the Internet. When a new movie or a video game is released, there is a rapid increase in the queries for its characters, scenes, and wallpapers on images search. For instance, when the *Avengers* movie was released, a large number of queries were issued for characters from this movie (e.g., Iron Man and the Incredible Hulk). Many image search engines employ the presence of faces to re-rank the retrieved images to better serve such queries. These systems need to be generic enough to be able to detect faces across huge variations—some of which may be non-human.[4] On one hand, it is challenging to build a single system that can detect these different human-like faces with such diverse appearances. On the other hand, it is infeasible to employ separate detectors for individual characters due to the large number of labeled examples and the large computation time required to train these detectors.

In addition to evaluate the proposed cascade adaptation technique on the images of faces of babies, Jain and Farfade [19] considered the images for different movie characters that are "human like." This *Human-like* dataset is collected by retrieving the top 1000 results for appropriate queries to the Bing image search engine. Figure 13 shows the detection results on images from these datasets. The respective performance curves show a clear improvement in performance obtained using the adapted cascade. A cascade trained from scratch using the training examples correspond to

[4]In fact, the main characters in six of the top 10 highest-grossing hollywood movies of the year 2012 are non-human characters that have appearances with strong similarity to humans.

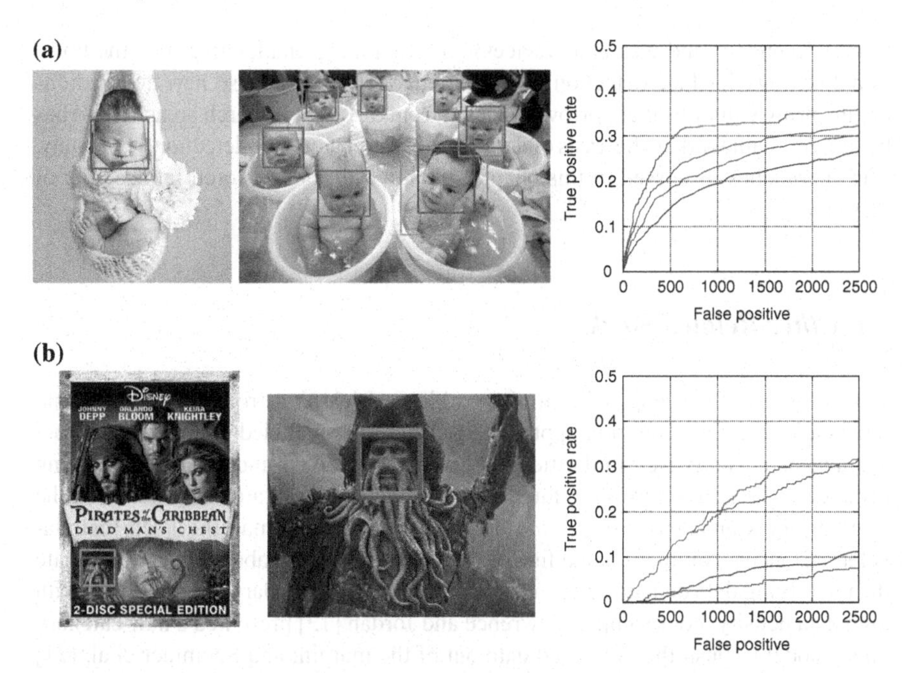


**Fig. 13** (Cols 1–2) Ground truth (*red*) and the output of our detector (*green*) are shown. Performing at the same false positive rate, the original cascade did not detect any of these face regions. (Col 3) Performance curves using the FDDB discrete matching score [17]: original cascade (*black*), original + validation (*magenta*), original + adaptation (*blue*), and original + adaptation + validation (*green*). The standard techniques for post-processing will improve the performance curves further uniformly for all of these approaches. However, since the objective is to study the effectiveness of the adaptation techniques, the focus is only on the algorithmic improvement in performance. **a** Baby photos. **b** Davy Jones (pirates of the carribean)

the first few stages of the cascade. Since these stages only serve the purpose of easy rejection, tens of thousands of false positives were observed for this cascade; the true positive rates are close to zero for <5 K false positives. These observations validate the efficacy of the above technique for adapting a pre-trained classifier to work well in a new domain. Moreover, this approach constructed an effective detector for this domain in a short period of time, whereas training similar detector from scratch would typically require several days of effort for annotation and model training.

## 5 Discussion

This chapter shows that simple adaptation techniques can substantially improve the performance of a black-box classifier in a new domain. Since both of the discussed approaches are computationally efficient, they are relevant to several new domains and applications. The settings for these new applications (particularly those relevant to mobile phones and wearable devices) are often more challenging than the traditional settings for face detection. The challenges posed by these new applications are inspiring solutions that borrow concepts from different research areas. There has been significant work in the areas of domain adaptation and face detection, separately. The approaches discussed in this chapter lie at a rather unexplored intersection of these areas.

## 5.1 Other Related Work

*Semi-supervised learning* refers to the problem of learning from both labeled and unlabeled data. One common approach for handling unlabeled data is to construct a graph using pairwise similarities between both labeled and unlabeled training instances. The nodes corresponding to the labeled instances are annotated with the original labels, and *label propagation* [14] is performed to estimate the labels for unlabeled instances. Another related line of research uses the unlabeled data to estimate the underlying data density and move the classification boundary out of regions with high data density. For example, Lawrence and Jordan [13] presented a null-category noise model to push the unlabeled data out of the margin; and Szummer et al. [21] included a regularization term based on a local estimate of the mutual information between the data and the label distributions to move the classification boundary out of the high-density data regions. The latter approach of *information regularization* was extended by Corduneanu et al. [11] for semi-supervised learning. The first approach presented in this chapter uses the similarity between the data points to update the detection score of the data points for which the predicted score from the pre-trained detector is near the classification boundary. This update effectively makes the data distribution sparse around the original classification boundary. While the intuition behind this procedure is similar to those of the above-mentioned methods, these semi-supervised learning approaches assume that both of the labeled and unlabeled data are sampled from an identical underlying distribution. This assumption does not hold true for several practical settings.

The problem formulations used in this chapter are also similar to the work in *domain adaptation*. In domain adaptation, labeled data from one or multiple "source" domains is used to train models to perform well on a different yet related "target" domain. Daumé and Marcu [12] approach this problem by modeling the data distribution for each of these domains as a mixture of a global- and a domain-specific component. This global component is inferred from the data of the source domain(s) and applied to the data of the target domain. Most of the works in domain adaptation suggest minimizing a convex combination of source and target empirical risk [10]. Thus the classifier needs to be re-trained (repeatedly) from scratch for every new domain. For face detection, the distribution of face appearances may vary significantly from one image to another or across similar domains. For single-image adaptation, the existing techniques for domain adaptation are prohibitively slow. Also, these techniques are not optimized for cascades of classifiers.

Both of the discussed adaptation methods could also be interpreted as regularizing the output of a face detection algorithm on the data manifold. Belkin et al. [22] proposed smoothing the discriminative function by controlling the complexity of the learned classifier through the norm of the desired function in the corresponding reproducing kernel Hilbert spaces. They further showed that this *manifold regularization* framework generalizes a large set of learning algorithms including ridge regression and support vector machines. This framework provides useful insights into the relation between the hypotheses for the original detector and the adapted detector, but there is a significant cost associated with re-training the classifier for a new domain. A similar argument holds true for the relevance of the previous work related to the analysis of *covariate shift* [23].

## References

1. Yang, M., Wu, Y., Crenshaw, J., Augustine, B., Mareachen, R.: Face detection for automatic exposure control in handheld camera. In: International Conference on Computer Vision Systems (2006)
2. Froba, B., Ernst, A.: Fast frontal-view face detection using a multi-path decision tree. In: Proceedings of Audio- and Video-Based Biometric Person Authentication (2003)
3. Pham, M.T., Cham, T.J.: Fast training and selection of Haar features using statistics in boosting-based face detection. In: International Conference on Computer Vision (2007)
4. Osadchy, M., LeCun, Y., Miller, M.L.: Synergistic face detection and pose estimation with energy-based models. J. Mach. Learn. Res. **8**, 1197–1215 (2007)
5. Jain, V.: Using context to enhance the understanding of face images. Ph.D. thesis, University of Massachusetts Amherst (2010)
6. Viola, P.A., Jones, M.J.: Robust real-time face detection. Int. J. Comput. Vis. **57**(2), 137–154 (2004)
7. Zhang, C., Zhang, Z.: A survey of recent advances in face detection. Technical report, Microsoft Research (2010)
8. Rowley, H.A., Baluja, S., Kanade, T.: Neural network-based face detection. IEEE Trans. Pattern Anal. Mach. Intell. (1998)
9. Schneiderman, H., Kanade, T.: Probabilistic modeling of local appearance and spatial relationships for object recognition. In: IEEE Conference on Computer Vision and Pattern Recognition (1998)
10. Blitzer, J., Crammer, K., Kulesza, A., Pereira, F., Wortman, J.: Learning bounds for domain adaptation. In: Advances in Neural Information Processing Systems (2008)
11. Corduneanu, A., Jaakkola, T.: On information regularization. In: Conference on Uncertainty in Artificial Intelligence (2003)
12. Daumé III, H., Marcu, D.: Domain adaptation for statistical classifiers. J. Artif. Intell. Res. **26**, 101–126 (2006)
13. Lawrence, N.D., Jordan, M.I.: Semi-supervised learning via Gaussian processes. In: Advances in Neural Information Processing Systems (2005)
14. Zhu, X.: Semi-supervised learning with graphs. Ph.D. thesis, Carnegie Mellon University (2005)
15. Rasmussen, C.E., Williams, C.K.I.: Gaussian Processes for Machine Learning. The MIT Press (2005)
16. Jain, V., Learned-Miller, E.: Online domain adaptation of a pre-trained cascade of classifiers. In: IEEE Conference on Computer Vision and Pattern Recognition (2011)

17. Jain, V., Learned-Miller, E.: FDDB: a benchmark for face detection in unconstrained settings. Technical report, University of Massachusetts Amherst (2010)
18. Mikolajczyk, K., Schmid, C., Zisserman, A.: Human detection based on a probabilistic assembly of robust part detectors. In: European Conference on Computer Vision, pp. 69–82 (2004)
19. Jain, V., Farfade, S.S.: Adapting classification cascades to new domains. In: International Conference on Computer Vision (2013)
20. Lee, D.D., Seung, H.S.: Algorithms for non-negative matrix factorization. In: Advances in Neural Information Processing Systems (2000)
21. Szummer, M., Jaakkola, T.: Information regularization with partially labelled datas. In: Advances in Neural Information Processing Systems (2002)
22. Belkin, M., Niyogi, P., Sindhwani, V.: Manifold regularization: A geometric framework for learning from labeled and unlabeled examples. J. Mach. Learn. Res. **7**, 2399–2434 (2006)
23. Bickel, S., Brückner, M., Scheffer, T.: Discriminative learning under covariate shift. J. Mach. Learn. Res. **10**, 2137–2155 (2009)

# Genetically Modified Logistic Regression with Radial Basis Function for Robust Software Effort Prediction

**Manas Gaur**

**Abstract** Existing logistic regression technique applies log-based function on the dataset and provides the necessary result for analysis. It has been for years for prediction analytics. This project develops a genetically modified evolutionary logistic regression technique using radial basis functions (GLR-RBF). This methodology employs three crucial stages. First stage is an evolutionary stage employing two fitness functions. Initially, the dataset obtained from the repository is modified using Genetic Algorithm (GA), followed by creation and simulation of radial basis neural network (RBFNN). The RBFNN is iteratively trained and simulated. The weight matrix generated in each stage is tested with a fitness function, which define the rule for selection of the best individual that will be appended to the dataset. The process of appending the weight matric to the dataset marks the second stage. In the third stage, the new set of covariates is fed into logistic regression operator which classifies the attributes based on effort response variable. This method has been tested in field of remote sensing image classification and other large dataset. This project also tests this technique for software effort prediction, which is regarded as the crucial activity before the commencement of the software development cycle. The model GLR-RBF was found be competitive when compared with similar standard models. The measure obtained from AUC indicates that GLR-RBF has reached the state-of-the-art.

**Keywords** Artificial neural network · Radial basis functions · Range-based classification · Logistic regression · Genetic algorithm · Decision tree and Bayesian classifier · Attribute reduction

## 1 Introduction

During the last few years [1], there has been tremendous rise in areas where pattern classification algorithms are applied. Examples of application come from medical diagnosis, handwritten character recognition, dynamic signature verification or

---

M. Gaur (✉)
M. Tech, Software Engineering, Delhi Technological University, Delhi, India
e-mail: manasgaur99@gmail.com



R. Singh et al. (eds.), *Machine Intelligence and Signal Processing*,
Advances in Intelligent Systems and Computing 390,
DOI 10.1007/978-81-322-2625-3_5

satellite image analysis and in many regards, classification is the most important statistical problem around.

To effectively design the software, we need accurate and precise estimation of the software resources. Machine learning is a concept of artificial intelligence applied to field of software engineering to enhance the estimation process of the software. In machine learning, trends in software project and correlation in data are identified, and there is identification of important attributes that majorly affect the software cost and effort. This work presents a competitive study of multiclass learning which combines statistical logistic regression with radial basis kernel using evolutionary method to optimize the input parameters for effective software cost estimation. Radial basis function is a non-linear method used to solve a non-linear problem set that is the software effort data. Before applying the approach, the dataset is refined using Genetic Algorithm (GA) which sorts out the attributes that have strong influence on the response variable. The approach GLR-RBF uses three-step process: firstly, the attributes GA is applied on the attribute to calculate influence parameter, followed by training of RBFNN and finally statistical logistic regression which generates confusion matrix which defines the efficiency of the approach. The approach of binomial logistic regression is not new [1, 2], and it has been applied to the training of radial basis function neural network [3]. However, it is not usually employed in machine learning literature. The method of logistic regression poses problems which applied to real classification problems, where we cannot assume that covariates will be additive and purely linear. Our approach overcome these difficulties by augmenting the input vector with new RBF variables. The work is carried out on the dataset from UCI, PROMISE software engineering repository.

The results obtained were compared with that of linear regression and Bayesian classification which are regarded as the traditional statistical approaches to pattern recognition. The main approach of GLR-RBF in multiclass learning is combining different statistical and soft computing elements such as Binomial Logistic regression, radial basis function and genetic algorithm [6]. The methodology is a step towards application of hybrid algorithms which works on the rule of detection of correlation in the dataset and reduction of dataset to enhance the basic statistical algorithm efficiency.

## 2 Literature Survey

This section presents an overview of some research related to the model created in this project. A perfect implementation of RBFNN depends on the number of neurons, spread values and proper fitness function to evaluate the weight in each iteration. There have been many learning algorithms modifying RBFNN in various ways. The first category simply places one RBF at each sample [10]. If all training samples are selected as hidden neurons, then the resultant RBF provides poor analytics and many noised samples will be created and resultant confusion matrix provides poor results. Other attempt in creation of RBFNN focuses on reducing the number of RBFs in the model [11]. The impact of creating a RBF network on the basis of number of neurons is to improve the efficiency of the learning process. Many approaches exist to determine the hidden neurons. For instance, the number and position of the RBFs may be fixed ad defined a priori [12], or differ by the clustering algorithm employed: k-means clustering and fuzzy k-means clustering [12].

An interesting alternative is to evolve RBFNN using evolutionary algorithms (EAs). A very complete state-of-the-art of the different approaches and characteristics of a wide range of EA and RBFNN combination is given in [13]. RBF network can be improved by the change in centres and width using genetic algorithm. This has been stated and defined in [5] and proved that it lower the prediction error by 0.05–0.07. The approach of software cost estimation using neural network has been addressed in [9] and modifies using input sensitivity analysis spread factor.

The application of Logistic Regression (LR) technique employed in this project mainly focuses on predicting faults in software quality as stated in [13], testing the efficiency of test cases and improving the test cases and finally categorizing the outcomes of the test cases that have to be studied and testing in [4, 12]. LR finds its application in categorizing the projects based on their cost. This has been successful using historical data of past project which have trained LR model that has been described in [6]. The existence of these applications has motivated us to create a model using evolutionary algorithm to train RBFNN using minimum set of hidden neurons and also create close range so that better classification can be done. An attribute reduction method is proposed based on GA with heuristic information. It separates approximate core attributes from the whole attributes set, and then represents the rest attributes with a group of genetic chromosomes using binary encoding. This improves the local searching ability of GA in the process of global optimizing. Furthermore, the method designs the fitness function that prefers finding shorter range which increases the classification accuracy on new data. Experiments of reduction and classification with developed method are conducted [4]. A dataset consists of a large number of attributes which are irrelevant and redundant with respect to classification. The presence of such attributes hinders the accuracy of the classifier, and therefore it is necessary to remove these attributes before passing them into the classifier. Paper puts forward a method which determine the influence of independent attributes on dependent attribute by GA. This process increases convergence rate and also improves the accuracy of the classifier.

Radial-based functions have been used in various patterns of recognition and function approximation problems. The spread coefficient is the most important parameter of the radial basis learning and can speed up the curve fitting process [10]. Logistic regression in software cost estimation provides realistic interval predictions where the cost will lie. Log-likelihood function is more convenient to work with as it is a monotonically increasing function, and the logarithm of a function achieves its maximum value at the same points as the function itself. Finding the maximum of a function often involved taking the derivative of a function and solving for the parameter being maximized and this is often easier when the function is log-likelihood [13]. The statistical techniques like decision tree and logistic regression are much similar in their process to analyse the dataset. The techniques find its application in various estimation problems like acute cardiac ischemia. Other techniques behaving in similar manner are Bayesian network in clinical practices.

# 3 Background

This section describes about the simple logistic regression technique which provided the motivation to develop and test the new evolutionary learning algorithm GLR-RBF which has been explained in subsequent subsection.

## *3.1 Logistic Regression (LR)*

Logistic Regression is a mathematical modelling approach in which the best fitting, yet least restrictive model, is desired to describe the relationship between several independent explanatory variables and a dependent dichotomous response variable. A straight line can be represented by only two parameters the slope (m) and the intercept (c). However, an S-shaped curve is much more complex and representing it parametrically is not so easy. It turns out that we transform the Y's to the logarithm of the odds of Y, then the transformed target variable is linearly related to X. If Y is an event and p is the probability of the event happening and 1 − p is the probability of not happening, then

$$odds = \frac{p}{1-p} \tag{1}$$

And it turns out that $linear\ predictors = \log(\frac{p}{1-p})$. We can write the model as

$$\log\left(\frac{p}{1-p}\right) = mX + c \tag{2}$$

From the data given, we know that X can compute p for each value of X. After this, the problem is essentially similar to linear regression. For the sigmoid curve, the variable needs to be transformed from the p-space to the Y-space.

### 3.2 GLR-RBF Learning Algorithm

The promising function in this algorithm is the RBFNN, which generates appropriate weights of attributes using non-linear radial basis function. The dataset is refined using GA so that the result through RBFNN is achieved with minimum error and minimum number of neurons. The application of two EA's in this approach makes it far better in terms of classification and prediction than linear regression, decision tree and Bayesian network. The algorithm for the approach is defined below.

1. After retrieving the dataset from the repository and dividing it into training, validation and test dataset, we apply GA for attribute reduction.
2. The fitness function defined for the dataset tries to show the impact of attribute on the response variable. This fitness function easily maps to the sigmoid curve which appositely represent the problem domain. The response variable is the effort (dependent variable), and attributes are the independent variable with 0.01 representing the constant for convergence.

   Fitness function value (Fval) is given as follows:

$$Fval = \frac{\log(attribute) - 0.01}{response\ variable} \tag{3}$$

3. The best individual attributes along with their values form the training set for RBFNN.
4. The algorithm iteration is set to 1000, with changing spread parameter, which defines the sensitivity of the trained network.
5. The creation and simulation of radial basis network require creation of target vectors. The target vector is defined as T= {0, 1}. The target vector has length equal to training dataset.
6. The weights in the network are generated using

$$B(x, w) = e^{\left(\frac{(\|x-c\|)^2}{r^2}\right)} \tag{4}$$

   where B(x, w) = {B1(x, $w_1$), B2(x, $w_2$), B3(x, $w_3$), B4(x, $w_4$), B5(x, $w_5$)………. $B_m$(x, $w_m$)}, and C is the weight of the connection between the $I$th input node and $J$th RBF.
7. The weight matrix W is estimated by means of evolutionary NN algorithm that optimizes the error function given by

$$L(W) = -\frac{1}{N}\sum_{n=1}^{N} log\ p(yn|xn, W) \tag{5}$$

8. Above equation defines odds ratio for a given attribute with its weight vector. It can be expanded as

$$L(W) = \frac{1}{N}\sum_{n=1}^{N}\left[-\sum_{l=1}^{j-1} yn * f(xn, Wl) + \log\sum_{l=1}^{j-1} e^{f(xn, Wl)}\right] \quad (6)$$

where n defines the number of attributes or features in the dataset and $l$defines the number of radial basis neurons in the hidden layer of the RBFNN. $Y_n$ defines the response variable which is transformed into binomial form that is 1 and 0. 1 defines the true value and 0 defines the false value.
9. If the dataset can be concretely classified as 1 and 0, then we can say that binominal classification can be done in logistic regression. If we cannot concretely classify that data into 1's and 0's, then we can surely create range-based classes on the response attribute.
10. After each iteration, the weights are generated using (4). These weights test (6) with the sole motive to minimize the value. The process of iteratively training the weights using RBFNN is stopped when the change in value of (6) is minute or maximum iterations have been achieved.
11. After this process, we take the UNION of the weight metric and initial covariates to form a bet set of data sample (covariates).
12. These sets of new covariates are fed into the LR operator along with GA-modified dataset. The result is demonstrated using confusion matrix.

### *3.3 Characteristic Features of the Algorithm*

1. Representation of the individuals: The algorithm evolves architecture and connection weights simultaneously, each individual being a fully specified RBFNN. Each connection is specified by a binary value, indicating whether the connection exists and a real value representing its weight.
2. Error and Fitness Functions: We consider (6) as a measure of fitness of weights in each iteration by RBFNN. We see g(x, $\beta$, W) as the function of weights in hidden layer and the bias on the output node.
3. Initialization of the population: the initial population is generated by trying to obtain RBFNN with the maximum possible fitness value. The initial sets of population are also weights generated with spread as large at 4.0.
4. Step-wise Improvement: the process of improving the weights of the attributes is by slow change in spread value which also improves the convergence rate.
5. Effect of Union Operation: The union operation improves the classification by generating a new set of covariates where the pattern recognition process is more enhanced and is observed via high accuracy rate achieved from logistic regression classifier.

**Table 1** Influence (fitness function) of the attributes on the dataset response variable effort

| S.no | Attributes | Fitness function value |
|---|---|---|
| 1. | ACAP | 8.649 |
| 2. | AEXP | 10.872 |
| 3. | MODP | 7.914 |
| 4. | TOOL | 7.712 |
| 5. | VEXP | 10.098 |
| 6. | LEXP | 9.346 |
| 7. | SCED | 12.373 |
| 8. | STOR | 10.675 |
| 9. | DATA | 9.332 |
| 10. | TIME | 9.332 |
| 11. | TURN | 9.332 |
| 12. | VIRT | 9.654 |
| 13. | CPLX | 9.654 |
| 14. | RELY | 8.875 |
| 15. | LOC | 1.362 |
| 16. | PCAP | 8.134 |

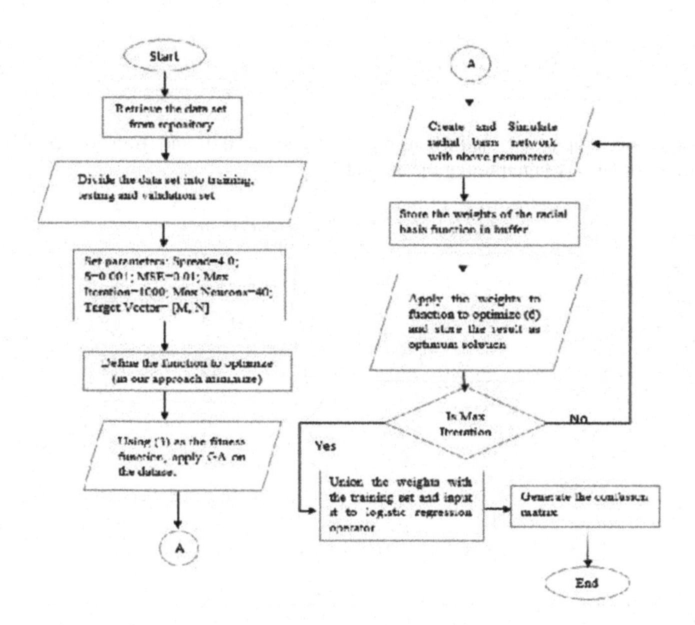


**Fig. 1** Flow chart of GLR-RBF

(a)

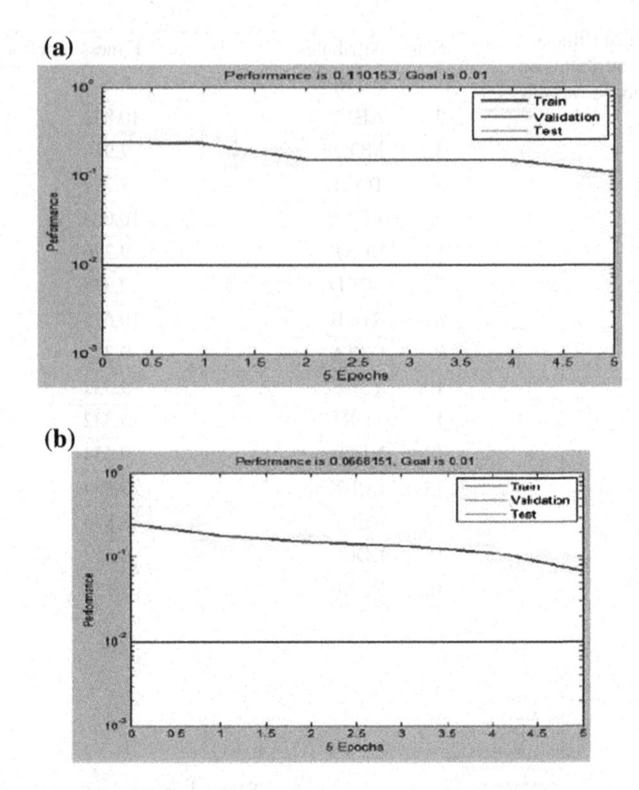


(b)

**Fig. 2** Performance graph generated when spread gradient (**a**) 0.01 and (**b**) 0.001. Goal 0.01 is the Mean square error taken when training the dataset

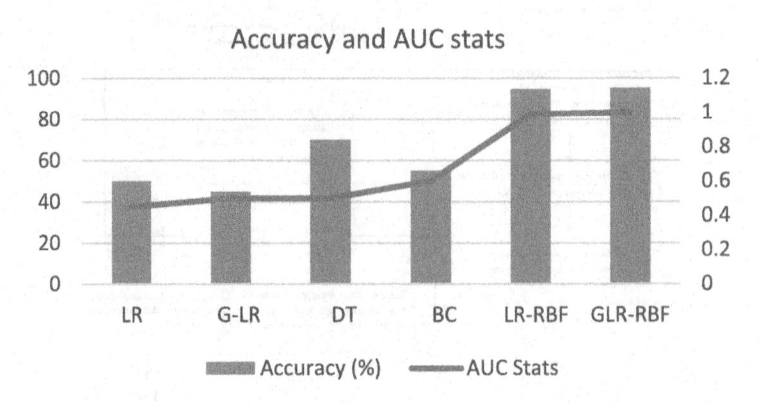


**Fig. 3** AUC Statistics

## 4 Results and Analysis

This section provides tabular comparative results when the developed technique is tested against other known statistical algorithms using real-time software data. The dataset used COCOMO81 to derive effort from multipliers. This dataset has 16 independent attributes and 1 dependent attribute (response variable), that is, effort. The standard machine learning techniques put to test in this work are Bayesian Classifier and Decision Tree.

Fitness function value in Table 1 is calculated using (3). In this process, each attribute's impact on the class label attribute effort is calculated, and a suitable constraint is applied to select the best attributes that have strong influence on the effort. The fitness values in the range of 9.3–11 are selected as attributes of high influence, as Log (base 10) values in these ranges are close to 10 and provide classification result similar to that obtained by the attributes in the original dataset. Original dataset has 16 attributes, but GA-modified dataset has 10 attributes.

Figure 2 clearly defines the essence of sensitivity in the creation and simulation of radial basis function. The small change in the spread value, the better train the RBFNN network. The weight matrix generated by the RBF network is tested using (6). If the generated value in (6) is less than previous, then an update of weight matrix is done. This process is repeated till the stopping criteria is met as explained in GLR-RBF learning algorithm and Fig. 1.

**Table 2** Result—Simple Logistic Regression on standard dataset (LR)

| | True range1 (0, 241.5] | True range2 (241.5, $\infty$) | Class precision |
|---|---|---|---|
| Predict range1 | 5 | 4 | 55.56 % |
| Predict range2 | 5 | 5 | 50 % |
| Class recall | 50 % | 55.56 % | |

**Table 3** Result—Simple Logistic Regression on Genetically modified dataset (G-LR)

| | True range1 (0, 241.5] | True range2 (241.5, $\infty$) | Class precision |
|---|---|---|---|
| Predict range1 | 5 | 5 | 50 % |
| Predict range2 | 5 | 4 | 44.44 % |
| Class recall | 50 % | 44.44 % | |

**Table 4** Result—Decision Tree Classifier on Standard dataset (DT)

| | True range1 (0, 241.5] | True range2 (241.5, $\infty$) | Class precision |
|---|---|---|---|
| Predict range1 | 8 | 3 | 72.73 % |
| Predict range2 | 2 | 6 | 75 % |
| Class recall | 80 % | 66.67 % | |

**Table 5** Result—Bayesian classifier on standard dataset (BC)

| | True range1 (0, 241.5] | True range2 (241.5, ∞) | Class precision |
|---|---|---|---|
| Predict range1 | 4 | 2 | 66.67 % |
| Predict range2 | 6 | 7 | 53.85 % |
| Class recall | 40 % | 77.78 % | |

**Table 6** Result—Logistic Regression-RBF on standard dataset (LR-RBF)

| | True range1 (0, 241.5] | True range2 (241.5, ∞) | Class precision |
|---|---|---|---|
| Predict range1 | 10 | 1 | 90.91 % |
| Predict range2 | 0 | 8 | 100 % |
| Class recall | 100 % | 88.89 % | |

**Table 7** Result—GLR-RBF on GA-modified dataset (GLR-RBF)

| | True range1 (0, 0.967] | True range2 (0.967, ∞) | Class precision |
|---|---|---|---|
| Predict range1 | 10 | 1 | 91 % |
| Predict range2 | 0 | 9 | 100 % |
| Class recall | 100 % | 91 % | |

**Table 8** Performance metric-based comparison of GLR-RBF with machine learning models

| S.no | Model | Accuracy (%) | AUC Stats |
|---|---|---|---|
| 1. | LR | 50 | 0.450 |
| 2. | G-LR | 45 | 0.500 |
| 3. | DT | 70 | 0.500 |
| 4. | BC | 55 | 0.600 |
| 5. | LR-RBF | 95 | 0.989 |
| 6. | GLR-RBF | 96 | 1.000 |

The quantitative results shown by confusion matrix from Tables 2, 3, 4, 5, 6, 7 and 8 prove that the technique GLR-RBF is much more robust and efficient in terms of binomial classification, especially in problem domain where the response variable can be easily be categorized into range-based classes. It can be further be proved from the study of Fig. 3, in which AUC statistics curve defines that accuracy of prediction and area under receiver operating characteristic curve (AUROC) of GLR-RBF is best when compared with DT, LR, and LR-RBF. The prediction percentage of 80 % and AUROC =1of GLR-RBF states that the transformation of application of GLR-RBF from remote sensing domain to software testing domain has reached state of the art. The performance metric table describes GLR-RBF as perfect classifier based on AUC statistics and accuracy in classification (Fig. 3 and Table 8).

## 5 Conclusion and Future Work

This project combines three powerful domains in machine learning and statistic research: Logistic Regression, Evolutionary Algorithms and Radial Basis Neural Network. The approach carries out an adequate combination of the three elements to solve multiclass problems. Specifically, this new approach consists of binomial logistic regression model build on the union of initial covariate and Gaussian RBFs for the predictor function. The process is carried out in three stages: firstly, dataset retrieved from the repository was genetically modified and fed into RBFNN, which is trained and validated, after which the new set of RBF and weight matrices is generated using activation function (sigmoid). Weight matrix is tested using a fitness function and reiterated with performance constraint if not met or maximum iteration count is not achieved. In the final stage, the weight matrix is appended with the initial covariates (GA modified) to form a new set of covariates with different ranges. This new dataset is provided as input to logistic regression operator. The results are obtained in the form of confusion matrix and AUC values. The results described in Sect. 4 define that GLR-RBF is a competitive method and has reached the state-of-the-art.

This GLR-RBF technique can be applied to different fields in software engineering for prediction, classification, improving the test suite, software refactoring and pattern recognition.

## References

1. Hervas-Martinez, C., Martinez-Estudillo, F.J., Carbonero-Ruz, M.: Multilogistic regression by means of evolutionary product-unit neural networks. Neural Netw. **14**, 201–208 (2004)
2. Krishnapuram, B., Carin, L., Figueiredo, M.A., Hartemink, A.J.: Sparse multinomial logistic regression: fast algorithm and generalization bounds. IEEE Trans. Pattern Anal. Mach. Intell. **27**, 957–968 (2005)
3. Nabney, I.T.: Efficient training of RBF networks for classification. Int. J. Neural Syst. **14**, 201–208 (2004)
4. Whitehead, B.A., Choate, T.D.: Cooperative-competitive genetic evolution of radial basis function centres and widths for time series prediction. In: IEEE Trans. Neural Netw. **7**, 869–880 (1996)
5. Sentas P., Angelis L., Stamelos, I.: Multinomial logistic regression applied on software productivity prediction. In: Proceedings of the 9th PanHellenic Conference in informatics (2003)
6. Salem, A.M., Rekab, K., Whittaker, J.A.: Prediction of software failures through logistic regression. Inf. Softw. Technol. Elsevier **46**, 781–789 (2004)
7. Sentas, P., Angelis, L., Stamelos, I.: Software productivity and effort prediction with ordinal regression. Inf. Softw. Technol. Elsevier **45**, 17–29 (2005)
8. Krishnamurthy, S., Fisher, D.: Machine learning approaches to estimating software development effort. IEEE Trans. Softw. Eng. 126–137 (1995)
9. Papatheocharous, E.: Software cost modelling and estimation using artificial neural network enhanced input sensitivity analysis. J. Univ. Comput. Sci. 1–30 (2012)
10. Orr, M.J.: Optimising the widths of radial basis functions. In: Proceedings of the 5th Bazillian Symposium on Neural Network, pp. 26–29 (1998)

11. Chen, S., et al.: Orthogonal least square learning algorithm for radial basis function. In: IEEE Trans. Neural Netw. **2**, 302–309 (1991)
12. Hunt K.J. et al.: Neural network for control system-A survey. In: Automatica **28**, 1083–1112 (1992)
13. Buchtala, O., Klimek M., Sick, B.: Evolutionary optimization of radial basis function classifier for data mining application. In: IEEE Trans. Syst. Man, Cybern., B; Cybern. **35**, 928–947 (2005)

# Recovering Partially Sampled EEG Signals Using Learned Dictionaries

**Arpita Gang, Janki Mehta and Angshul Majumdar**

**Abstract** Prior studies have proposed methods to recover multi-channel EEG signal ensembles from their partially sampled entries. These techniques are only limited to multi-channel scenarios, especially when the number of channels is very large. Most biomedical signals (apart from EEG) are acquired from a single channel or from very few channels. Existing techniques cannot be employed to recover such signals from their partial samples. In this work, we propose a dictionary learning-based technique to overcome the problem. A sparsifying dictionary is learnt from the training examples; the trained dictionary is then used as a sparsifying transform in compressed sensing settings to recover the partially sampled test signal. Our proposed dictionary learning-based technique shows significant improvement in recovery over fixed sparsifying basis.

**Keywords** Dictionary learning · Compressed sensing · Wireless body area network

## 1 Introduction

Recently, Wireless Body Area Networks (WBAN) are being used for monitoring the health of individuals remotely (at base station). Various types of health indicating signals such as EEG, ECG, Pulse Plethysmograph, etc. can be monitored thus. In this work, we are particularly interested in the acquisition and transmission of EEG

A. Gang (✉) · J. Mehta · A. Majumdar
Indraprastha Institute of Information Technology Delhi,
New Delhi 110020, India
e-mail: arpita1478@iiitd.ac.in
URL: http://www.iiitd.ac.in/

J. Mehta
e-mail: mehta1485@iiitd.ac.in

A. Majumdar
e-mail: angshul@iiitd.ac.in



R. Singh et al. (eds.), *Machine Intelligence and Signal Processing*,
Advances in Intelligent Systems and Computing 390,
DOI 10.1007/978-81-322-2625-3_6

signals over WBAN. EEG signals are useful for detecting epileptic seizures and strokes.

The main problem in any wireless sensor network (including WBAN) is its limited energy resources. The main computational power for any sensor network is expended in sampling, processing and transmitting. The transmission stage is the most power consuming, followed by sampling. The cost of processing is small compared to the other two operations. To save on transmission power, the signal must be compressed.

In traditional source coding (for compression), the encoder is complex and the decoder is simple. However, in this scenario, the requirement is exactly the opposite. The encoder needs to be simple (less power consuming) since this operation is carried out at the nodes. The decoder is at the base station where there is no constraint on computational resources.

Compressed Sensing (CS)-based encoder–decoder is ideal for this scenario. CS uses random projections (from higher to lower dimensions) for encoding; this operation is computationally cheap. Compression enables saving of transmission power. CS decoding is computationally expensive and requires solving a non-linear optimization problem. However, as mentioned before, this poses no issue since there are powerful computers at the base station. Such CS-based encoding–decoding was first proposed in [1] and has been subsequently used in [2–6].

Some recent studies have devised techniques to reduce the acquisition power by subsampling the signal [7, 8]. Here, the multi-channel EEG signal ensemble is recovered using sparse and low-rank techniques. Partial sampling can drastically reduce the power consumption for data acquisition and processing. It has been shown in [7, 8] that for 20 % under-sampling, the overall (acquisition + processing + transmission) energy saving is over 50 and for 40 % under-sampling, it is over 30 %.

Existing CS-based single-channel recovery techniques [1–6] cannot be used to recover partially sampled signals. Blind Compressed Sensing (BCS) [7] and low-rank [8]-based methods can only recover partially sampled signals for multi-channel ensembles, especially when the number of channels is high. This is practical for EEG signals since they are typically collected over 32 or more channels, but not for other biomedical signals like ECG or Pulse Plethysmograph.

In this work, we propose a new technique that can recover partially sampled biomedical signals that are acquired by single channel. The main challenge in employing CS-based recovery techniques is to satisfy the incoherence condition, and the sparsifying dictionary needs to be maximally incoherent with the Dirac sampling basis. None of the commonly used sparsifying transforms (wavelet, DCT, Gabor, etc.) satisfy this condition [9]; hence, CS recovery results using these transforms are poor.

We propose to circumvent this issue by learning the sparsifying dictionary empirically. K-SVD [12] is used to train the sparsifying dictionary from training data; the learned dictionary is used to recover subsampled signals. The rest of the paper is organized into several sections. Prior art is reviewed in Sect. 2. The proposed technique is described in Sect. 3. The basic tenets of dictionary learning are discussed in Sect. 3.1. We discuss the experimental results in Sect. 4. Finally, the conclusions of this work and scope of further research are discussed in Sect. 5.

## 2 Literature Review

One of the earliest works that applied CS for EEG signal compression and transmission is [1]. It projected the EEG signal onto an IID Gaussian basis for compression and used CS to recover the EEG signal by exploiting the signals sparsity in the Gabor domain. The work [1] employed a synthesis prior formulation for sparse signal recovery using Gabor as a sparsifying basis. Synthesis prior only works for orthogonal and tight-frame sparsifying transforms such as DCT, wavelets, curvelets, etc. Unfortunately, Gabor is not an orthogonal or tight-frame transform; therefore, their proposal to recover EEG signals, thus, is theoretically incorrect. Also, the said paper uses a Gaussian matrix for compression; for practical reasons, a Gaussian compression basis is not very suitable since the matrix is dense, and therefore is neither memory efficient nor easy to operate with.

In [2], different sparsifying transforms were compared wavelets, Gabor, and splines; it was reported that Gabor yielded the best reconstruction results. However, this paper is fraught with the same issue as [1]; they pose EEG reconstruction as a synthesis prior problem using the Gabor basis albeit incorrectly.

The possibility of exploiting inter-channel correlation in order to improve EEG signal reconstruction was mentioned in [1], but there was no concrete idea regarding how to model it. This problem is partially addressed in [3]. They do not explicitly model the inter-channel correlation, but frame a joint reconstruction problem where the signals from all the channels are reconstructed simultaneously. This work uses wavelets as the sparsifying basis and a binary matrix of randomly positioned 1's and 0's as the compression basis. Such a binary compression matrix is easy to store and operate with.

A more recent work model assumes a block structure of the EEG signals [4] in a transform domain (DCT or wavelet). However, there is no theoretical or physical intuition behind this assumption; but it was shown in [8] that a Block Sparse Bayesian Learning (BSBL) algorithm yields good recovery results. This work too uses a random binary matrix for projection.

Studies like [1, 2] posed EEG recovery as a synthesis prior problem albeit incorrectly. This issue is corrected in [4] by posing reconstruction as an analysis prior problem.

The work [3] claims to exploit inter-channel correlations for EEG signal recovery, but in essence does not. It only uses the sparsity of the EEG signals in the wavelet domain and such unstructured sparse recovery problems do not exploit inter-channel correlations. In [4], it was shown for the first time that inter-channel correlations can be cast as a group-sparse recovery problem.

The studies discussed so far [1–6] compressed the EEG signal after full sampling. This reduces the transmission energy only. In [7, 8], it was shown that how acquisition and processing energy can be reduced by partially subsampling the EEG signals; the reconstruction was carried out by exploiting the intra-channel and inter-channel correlations. In [7], the recovery was posed as a low-rank matrix completion problem. In [8], the Blind Compressed Sensing (BCS) framework was used to solve the multichannel EEG recovery problem.

## 3 Proposed Formulation

EEG signals are almost always acquired via multiple channels. Therefore, the solutions proposed in [7, 8] suffice. However, for most other biomedical signals, ECG, Pulse Plethysmograph, etc., the acquisition is via a single channel or very few channels. In such cases, the solutions proposed in [7, 8] are not applicable. But there is still a need to recover these signals from their partially sampled entries. The subsampling operation is expressed as

$$\boldsymbol{b} = \boldsymbol{R}\boldsymbol{x} + \eta \tag{1}$$

Here, $\mathbf{x}$ is the underlying signal, $\mathbf{b}$ is the partially sampled signal, $\mathbf{R}$ is the sampling operator, and the noise $\eta$ is assumed to be Gaussian.

Synthesis prior to CS techniques will employ wavelets or DCT to sparsify the signal and propose recovery via

$$\min_{\boldsymbol{\alpha}} \| \boldsymbol{b} - \boldsymbol{R}\boldsymbol{\Phi}^T\boldsymbol{\alpha} \|_2^2 + \lambda \| \boldsymbol{\alpha} \|_1 : \boldsymbol{x} = \boldsymbol{\Phi}\boldsymbol{\alpha} \tag{2}$$

where $\boldsymbol{\Phi}$ is the wavelet/DCT operator and $\boldsymbol{\alpha}$ are the sparse transform coefficients.

Analysis prior techniques would use Gabor as the sparsifying basis and propose recovery as

$$\min_{\boldsymbol{x}} \| \boldsymbol{b} - \boldsymbol{R}\boldsymbol{x} \|_2^2 + \| \boldsymbol{H}\boldsymbol{x} \|_1 \tag{3}$$

where $\mathbf{H}$ is the Gabor basis.

Both of these formulations suffer from a fundamental limitation. Such CS recovery techniques require that the incoherence between the measurement basis (in our case Dirac) and the sparsifying transform (in our case wavelet/DCT/Gabor) should be high [9]; otherwise, the recovery will be poor. Unfortunately, this condition is not satisfied in the present scenario. Commonly used sparsifying transforms like wavelet, DCT, or Gabor are not very incoherent with the Dirac sampling basis; therefore, the recovery results using these bases are unsatisfactory.

To overcome this issue, we propose to learn the sparsifying dictionary empirically. Previously, such learned dictionaries have been used for signal denoising [10] and signal classification [11]. For the first time, in this work, we realize the need for learning empirical sparsifying dictionaries for signal reconstruction. We employ the learned dictionary to recover the signal using the synthesis prior formulation.

### 3.1 K-SVD Dictionary Learning [12]

K-SVD aims at finding the best dictionary that sparsifies the set of signals in $\mathbf{Y} = \{\mathbf{y}_i\}_{i=1}^N$ by solving

$$\min_{\boldsymbol{D},\boldsymbol{X}} \| \boldsymbol{Y} - \boldsymbol{D}\boldsymbol{X} \|_F^2 \text{ subject to } \forall i, \| \boldsymbol{x}_i \|_0 \leq T_0 \tag{4}$$

Sparse coding stage involves solving N optimization problems given by equation below using any pursuit algorithm for a fixed initial dictionary $\mathbf{D}^{(0)} \in \mathbb{R}^{n \times K}$ with $l^2$ normalized columns:

$$\min_{\boldsymbol{x}_i} \| \boldsymbol{y}_i - \boldsymbol{D}\boldsymbol{x}_i \|_2^2 \text{ subject to } \| \boldsymbol{x}_i \|_0 \leq T_0 \tag{5}$$

for $\forall i = 1, 2, \ldots, N$.

Next is the Dictionary Update Stage: Here, only one column of the dictionary is updated at a time keeping $\mathbf{X}$ fixed. For this, we rewrite the cost function as follows:

$$\begin{aligned} \| \boldsymbol{Y} - \boldsymbol{D}\boldsymbol{X} \|_F^2 &= \left\| \boldsymbol{Y} - \sum_{j=1}^{K} \boldsymbol{d}_j \boldsymbol{x}_j^T \right\|_F^2 \\ &= \left\| \left( \boldsymbol{Y} - \sum_{j \neq k} \boldsymbol{d}_j \boldsymbol{x}_j^T \right) - \boldsymbol{d}_k \boldsymbol{x}_k^T \right\|_F^2 \\ &= \| \boldsymbol{E}_k - \boldsymbol{d}_k \boldsymbol{x}_k^T \|_F^2 \end{aligned} \tag{6}$$

where $\mathbf{x}_k^T$ is $k^{th}$ row of $X$. The matrix $\mathbf{E}_k$ represents the error matrix for all the $N$ signals when the $k^{th}$ atom is removed.

Now, $\mathbf{x}_k^T$ is not necessarily sparse as the sparsity constraint is not applied. So we define

$$\omega_k = \{i | 1 \leq i \leq N, \boldsymbol{x}_k^T(i) \neq 0\} \tag{7}$$

These indices are such that signals $\{\mathbf{y}_i\}$ use atom $\mathbf{d}_k$ and so only they are significant while minimizing above cost function. Thus, we restrict our error matrix $\mathbf{E}_k$ to the columns corresponding to $\omega_k$, to get $\mathbf{E}_k^R$.

The SVD decomposition gives $\mathbf{E}_k^R = \boldsymbol{U}\boldsymbol{\Delta}\mathbf{V}^T$. Now $\mathbf{d}_k$ is updated to the first column of $\mathbf{U}$ and $\boldsymbol{x}_k^R$ is updated to the $\boldsymbol{\Delta}(1, 1)$ times the first column of $\mathbf{V}$. Taking these values ensures that $\boldsymbol{d}_k \boldsymbol{x}_k^R$ is the closest rank-1 matrix to $\mathbf{E}_k^R$, thus minimizing the cost function.

By this way, entire dictionary $\mathbf{D}$ is updated and then iteratively $\mathbf{X}$ and $\mathbf{D}$ are updated till a stopping criterion (usually number of iterations) is met.

## 4 Experimental Evaluation

The experiments are carried out on the BCI Competition III dataset 1 [13]. During the BCI experiment, a subject had to perform imagined movements of either the left small finger or the tongue. The time series of the electrical brain activity was

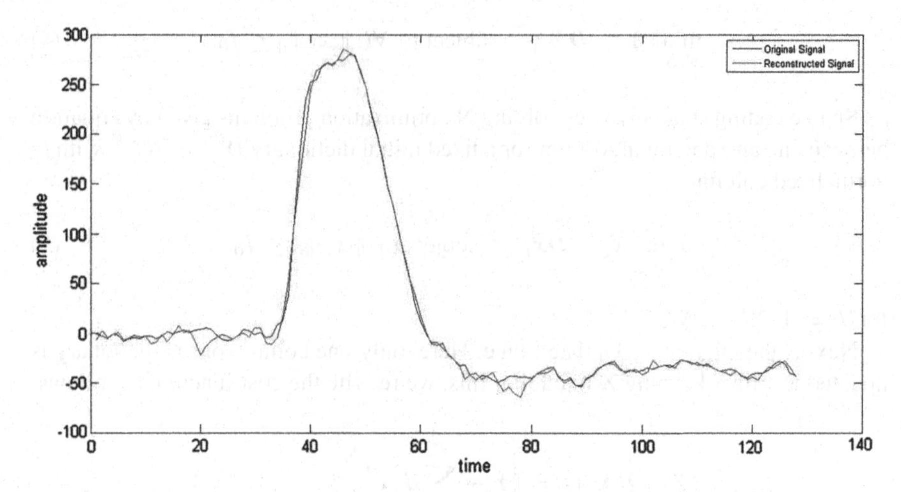


**Fig. 1** Overlaid K-SVD DL reconstruction on fully sampled ground truth (Channel 1)

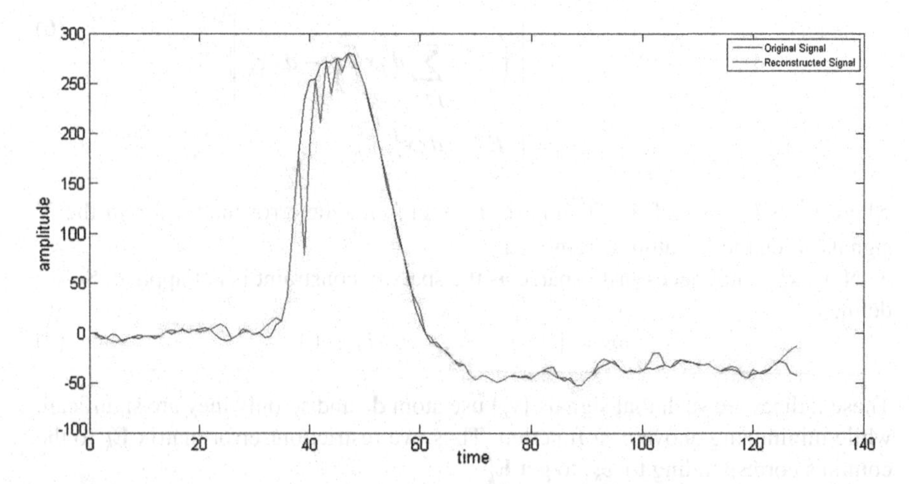


**Fig. 2** Overlaid wavelet reconstruction on fully sampled ground truth (Channel 1)

picked up during these trials using a 8 × 8 ECoG platinum electrode grid which was placed on the contralateral (right) motor cortex. The grid was assumed to cover the right motor cortex completely, but due to its size (approx. 8 × 8 cm), it partly covered also surrounding cortex areas. All recordings were performed with a sampling rate of 1000 Hz. After amplification, the recorded potentials were stored as microvolt values. Every trial consisted of either an imagined tongue or an imagined finger movement and was recorded for 3 s duration. To avoid visually evoked potentials being reflected by the data, the recording intervals started 0.5 s after the visual cue had ended.

The labeled training data consists of the brain activity during 278 trials. The test data consists of 100 trials. Since we are interested in single-channel reconstruction, we selected one channel and used the training set for that channel to learn the

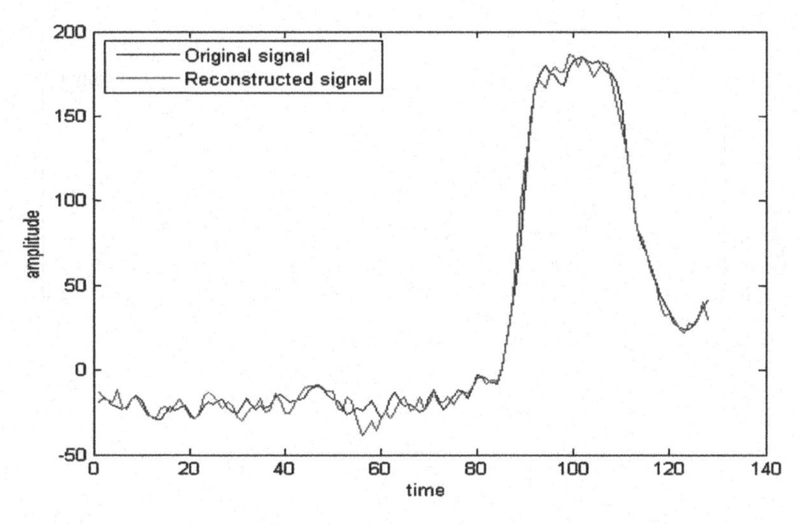


**Fig. 3** Overlaid K-SVD DL reconstruction on fully sampled ground truth (Channel 8)

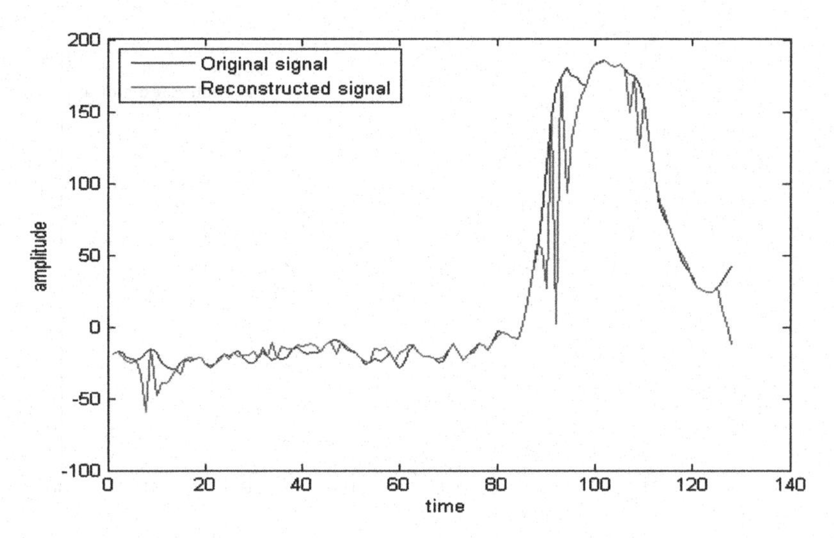


**Fig. 4** Overlaid wavelet reconstruction on fully sampled ground truth (Channel 8)

dictionary. The dictionary was learnt for signals of length 128 with a redundancy (over-completeness) of 2.

For evaluating the reconstruction results from partial sampling, the test dataset was used. The test signals were partially sampled with a sampling ratio of 50 %. We used the synthesis prior formulation for reconstructing the subsampled signals using the learned dictionary. We compare the learned dictionary with a wavelet dictionary (Daubechies wavelet, four vanishing moments, and five levels of decomposition). The normalized mean squared error between wavelet reconstruction and the ground truth is 0.2739, and for our proposed dictionary learning approach, it is 0.1308.

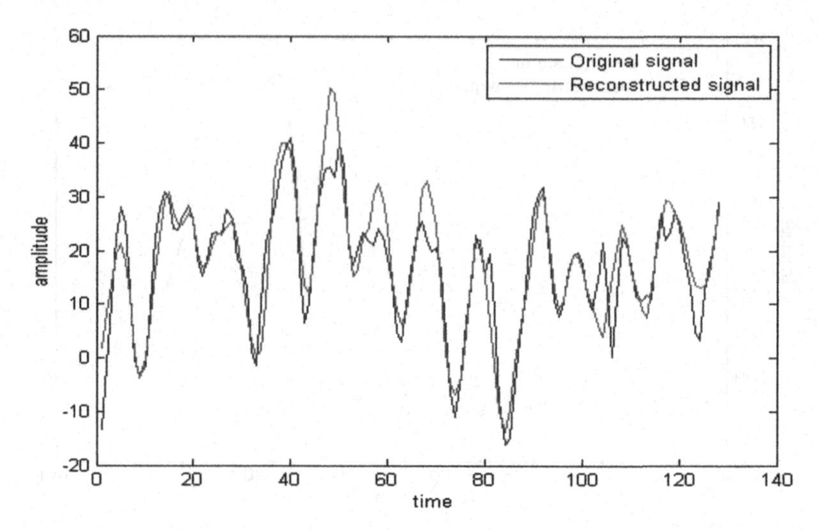


**Fig. 5** Overlaid K-SVD DL reconstruction on fully sampled ground truth (Channel 20)

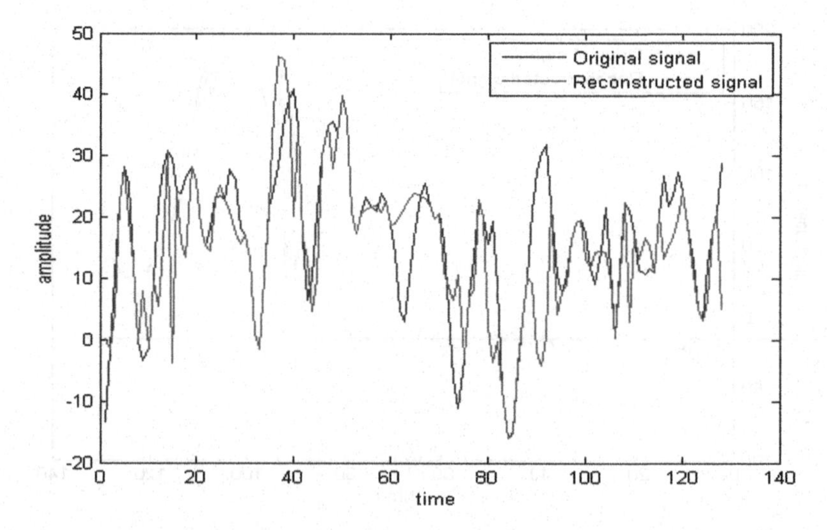


**Fig. 6** Overlaid wavelet reconstruction on fully sampled ground truth (Channel 20)

The results clearly show that the empirical dictionary yields significantly better results than the wavelet transform. For visual clarity, the reconstructed and the original signals are overlaid in the figures. Figures 1 and 2 show the difference in results of wavelet and learned dictionary for a signal from channel 1. Similarly, Figs. 3 and 4 show the reconstruction for channel 8, and Figs. 5 and 6 reflect the reconstruction for channel 20. The reconstruction from K-SVD dictionary learning almost matches the fully sampled ground truth. But the reconstruction from wavelet shows severe artifacts.

## 5 Conclusion

This work addresses the problem of recovering single-channel biomedical signals from their partial samples. Standard CS-based techniques based on fixed sparsifying dictionaries fail in this scenario (as shown in the experiments). To ameliorate this issue, we propose learning the sparsifying transform. We show that a simple K-SVD-based dictionary learning can improve the reconstruction results significantly and produce near-perfect recovery.

This work is a proof-of-concept. We argued that standard fixed dictionaries like wavelet, DCT, and Gabor are not incoherent with the Dirac sampling basis, and hence we need to learn the dictionary. We were fortunate that a dictionary learned using K-SVD yielded good results, but that need not be the case. K-SVD does not explicitly learn dictionaries that are incoherent with the measurement basis. However, there are more sophisticated dictionary learning techniques that can learn dictionaries with incoherence constraints [14, 15]. In the future, we will employ these techniques to solve our problem and expect to improve the results even further.

## References


1. Aviyente, S.: Compressed sensing framework for EEG compression. In: IEEE Workshop on Statistical Signal Processing, pp. 181–184 (2007)
2. Abdulghani, A.M., Casson, A.J., Rodriguez-Villegas, E.: Quantifying the performance of compressive sensing on scalp EEG signals. In: International Symposium on Applied Sciences in Biomedical and Communication Technologies, pp. 1–5; 7–10 (2010)
3. Kamal, M.H., Shoaran, M., Leblebici, Y., Schmid, A., Vandergheynst, P.: Compressive multichannel cortical signal recording. In: Proceedings of the 38th International Conference on Acoustics, Speech, and Signal Processing (ICASSP), Vancouver, Canada (2013)
4. Zhang, Z., Tzyy-Ping, J., Makeig, S., Rao, B.D.: Compressed sensing of EEG for wireless telemonitoring with low energy consumption and inexpensive hardware. IEEE Trans. Biomed. Eng. (accepted)
5. Mohsina, M., Majumdar, A.: Gabor based analysis prior formulation for EEG signal reconstruction. Biomed. Signal Process. Control **8**(6), 951955 (2013)
6. Majumdar, A., Ward, R.K.: Non-convex row-sparse MMV analysis prior formulation for EEG signal reconstruction. Biomed. Signal Process. Control **13**, 142147 (2014)
7. Shukla, A., Majumdar, A.: Row-sparse blind compressed sensing for reconstructing multichannel EEG signals. Biomed. Signal Process. Control (accepted)
8. Majumdar, A., Gogna, A., Ward, R.: Low-rank matrix recovery approach for energy efficient EEG acquisition for wireless body area network. Sens. Special Issue State-of-the-art Sens. Technol. Can. **14**(9), 15729–15748 (2014)
9. Cands, E., Romberg, J.: Sparsity and incoherence in compressive sampling. Inverse Prob. **23**(3), 969 (2007)
10. Hamner, B., Chavarriaga, R., Jos del, R.M.: Learning dictionaries of spatial and temporal EEG primitives for brain-computer interfaces. In: Workshop on Structured Sparsity: Learning and Inference, ICML (2011)
11. Zhou, W., Yang, Y., Yu, Z.: Discriminative dictionary learning for EEG signal classification in Brain-computer interface. In: Proceedings of the 12th International Conference on Control Automation Robotics and Vision (ICARCV), pp. 1582–1585 (2012)

12. Aharon, Michal, Elad, Michael, Bruckstein, Alfred: K-SVD: an algorithm for designing overcomplete dictionaries for sparse representation. Inverse Prob. **54**(11), 4311–4322 (2006)
13. BCI Competition III, http://www.bbci.de/competition/iii/
14. Arora, S., Ge, R., Moitra, A.: New algorithms for learning incoherent and overcomplete dictionaries. arXiv:1308.6273
15. Barchiesi, D., Plumbley, M.D.: Learning incoherent dictionaries for sparse approximation using iterative projections and rotations. IEEE Trans. Signal Process. **61**(8), 2055–2065 (2013)

# Retinal Vessel Classification Based on Maximization of Squared-Loss Mutual Information

**D. Relan, L. Ballerini, E. Trucco and T. MacGillivray**

**Abstract** The classification of retinal vessels into arterioles and venules is important for any automated system for the detection of vascular changes in the retina and for the discovery of biomarkers associated with systemic diseases such as diabetes, hypertension, and cardiovascular disease. We introduce Squared-loss Mutual Information clustering (SMIC) for classifying arterioles and venules in retinal images for the first time (to the best of our knowledge). We classified vessels from 70 fundus camera images using only 4 colour features in zone B (802 vessels) and in an extended zone (1,207 vessels). We achieved an accuracy of 90.67 and 87.66 % in zone B and the extended zone, respectively. We further validated our algorithm by classifying vessels in zone B from two publically available datasets—INSPIRE-AVR (483 vessels from 40 images) and DRIVE (171 vessels from 20 test images). The classification rates obtained on INSPIRE-AVR and DRIVE dataset were 87.6 and 86.2 %, respectively. We also present a technique to sort the unclassified vessels which remained unlabeled by the SMIC algorithm.

**Keywords** Retinal · Fundus · Vessels · Arterioles · Venules · Classification · Clustering

D. Relan (✉) · T. MacGillivray
Centre for Clinical Brain Sciences, Edinburgh, UK
e-mail: devanjali.relan@ed.ac.uk

L. Ballerini · E. Trucco
School of Computing, University of Dundee, Dundee, UK

T. MacGillivray
Clinical Research Imaging Centre, University of Edinburgh, Edinburgh, UK

T. MacGillivray
Clinical Research Facility, University of Edinburgh, Edinburgh, UK





R. Singh et al. (eds.), *Machine Intelligence and Signal Processing*,
Advances in Intelligent Systems and Computing 390,
DOI 10.1007/978-81-322-2625-3_7

## 1 Introduction

The retina is the only location in the human body where blood vessels can be directly visualized non-invasively in vivo. Many important eye diseases as well as systemic diseases such as diabetes and hypertension manifest themselves in the retina [1]. Quantitative structural analysis of the retinal vasculature helps in the diagnosis of retinopathies as well as providing candidate biomarkers of systemic diseases. The vascular changes during the onset of a systemic disease can affect very differently arteries and veins. For example, one of the early signs of retinopathy is generalized arteriolar narrowing [2], i.e. a decrease in the ratio between arteriolar and venular diameters (AVR). There is also mounting evidence that narrowed retinal arterioles are associated with long-term risk of hypertension, while AVR is a well-established predictor of stroke and other cardiovascular events [3, 4]. In order to realize an automatic tool such as VAMPIRE (Vasculature Assessment and Measurement Platform for Images of the REtina) [5] for vasculature characterization, it is highly desirable to automatically distinguish vessels as arterioles and venules.

Conventionally, AVR and other measurements of retinal vascular parameters are measured from zone B—an annulus 0.5 to 1 optic disc (OD) diameter from the OD boundary [2–4] and this region has been used for the investigation of retinal biomarkers in several studies [2, 4, 6]. However, measurements from outside this zone might also lead to candidate biomarkers of disease. For instance, Cheung et al. studied the association of blood pressure with retinal vascular caliber measured over the standard zone B and an extended zone of the fundus images [7]. They found that the reliability of retinal vascular caliber measurements was high for the extended zone.

We propose an automatic unsupervised retinal vessel clustering method based on Squared-loss Mutual Information (SMIC) [8] for the first time (to the best of our knowledge). We tested this on vessels in zone B and an extended zone using 4 fixed features frequently chosen or selected automatically as discriminative features for vessel classification [9].

Supervised vessel classification has previously been proposed in many studies [10–12], but this generally requires large volumes of clinical annotations on images. Therefore we attempt to classify vessels in unsupervised manner removing this burden. In our previous work [13], we proposed an unsupervised vessel classification method using a Gaussian Mixture Model with an Expectation-Maximization (GMM-EM) classifier. We demonstrated 92 % classification with 13.5 % unclassified vessels using a quadrant pair wise approach and 4 fixed features on 406 vessels from 35 images in zone B. We repeated the experiment on a further 35 images giving 70 images in total with 802 vessels and resulting in 90.45 % correct classification and with 13.8 % unclassified. The limitation of method was that it gave us a high percentage of unclassified vessels. In another study [14] we presented supervised a Least Square-Support Vector Machine (LS-SVM) classifier to label vessels automatically. The performance of our system was very promising and it only required a small set of training images. But like other supervised method our LS-SVM classification framework needs new training data for every new image set.

In [15], vessel classification was performed on 58 images using K-Means clustering between two concentric circumferences around the OD. In their study quadrants were rotated in steps of 20° to include at least one artery and one vein in each quadrant. They reported 87 and 90.08 % correct classification before and after applying their proposed tracking method in [16]. In [17], quadrant-wise vessel classification was performed using fuzzy a C-Mean clustering method in a concentric zone around the OD on 443 vessels from 35 images. Their proposed method resulted in 87.6 % correct classification. In [18], fuzzy C-Mean clustering was applied on 15 images for vessel classification with 88.28 % classification accuracy.

In many of the previous studies the INSPIRE-AVR [10] and DRIVE [19] datasets were used to test supervised classifiers. The authors in [10] tested on the INSPIRE-AVR dataset to classify vessels in zone B with 27 features, and their proposed method gave the best result with Linear Discriminant Analysis (LDA) (area under ROC 0.84). Mirsharif et al. [11] evaluated the performance of their supervised classification method on two different datasets viz. DRIVE (40 retinal images) and their own dataset (13 retinal images) resulting in 90.2 and 88.2 % accuracy, respectively. Dashtbozorg et al. in [12] presented a combination of graph-based classification with feature selection using LDA to classify all vessel pixels in entire images. Their method resulted in accuracy values of 88.3, 87.4 and 89.8 % with INSPIRE-AVR (40 images), DRIVE (20 images) and VICAVR (58 images) dataset, respectively.

In order to validate the performance of our system we therefore selected to test our algorithm on the INSPIRE-AVR and DRIVE datasets.

## 2 Methodology

### *2.1 Material*

70 colour fundus images were selected at random from a large database of non-mydriatic fundus images (Orkney Complex Disease Study—ORCADES). These images with resolution 2048 × 3072 pixels were captured with a Canon CR-DGi non- mydriatic retinal camera having 45° field of view. 802 vessels in zone B and 1,207 vessels in the extended zone were classified computationally. We also tested our algorithm on 171 vessels from 40 INSPIRE-AVR images and 483 vessels from 20 test DRIVE images.

Ground truth for the 70 ORCADES images was generated by two observers (authors DR and TM who were individually and independently trained by experienced clinicians in the identification of the retinal vessel type) by manual labelling vessels as arteriole or venule. Observer 1 (DR) is a doctoral student (less experienced, ~2.5 year); observer 2 (TM) is an imaging scientist and specialist of retinal image analysis (more experienced, >12 years).

## 2.2 Image Pre-processing

The hue channel was pre-processed to improve the contrast of vessels against background by mapping the original intensity values such that values between 0.01 and 0.8 map to values between 0 and 1. We then compensated for background illumination in red, green and re-mapped hue channels using Chrástek's method [20] as the presence of inter- and intra-image contrast, luminosity and colour variability affect classification performance.

## 2.3 Vessels and Features Extraction

Centerline pixels of vessels in zone B were extracted as previously reported [13], and vessels in the extended zone were extracted as reported in [14]. The image was divided into four quadrants by locating the OD and its diameter. Once the centerline pixels were found in each quadrant, four colour features—mean of red (MR), mean of green (MG), mean of hue (MH) and variance of red (VR)—were extracted from the illuminated corrected channels and from a circular neighborhood around each centerline pixel, with diameter 60 % of the mean vessel diameter.

After feature extraction, we have four sets of feature vectors $F_q, q = 1, \ldots, 4$, for each pair of quadrant—i.e. for pairs (I, II), (II, III), (III, IV) and (IV, I). Each set of colour features ($F_q$) of pixels were classified using the SMIC classifier. The classification was performed by working separately on two quadrants at a time in a clockwise direction, i.e. classification of vessels from quadrant combinations (I, II), (II, III), (III, IV) and (IV, I). The classifier labels each pixel as either arteriole or venule. Thereafter, labels were polled for all pixels in a vessel, and the winning vote used to assign the vessel status. If the vote was tied the vessel was marked unclassified. Vessels from ORCADES and DRIVE images were classified using four features (MR, MG, VR and MH) whereas INSPIRE-AVR images were classified using MR, MG and VR due to poor hue channel quality (see Fig. 1).

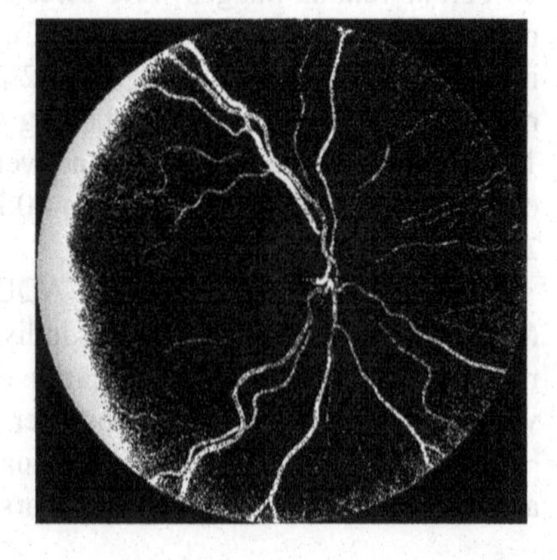

**Fig. 1** Poor hue channel of image from INSPIRE-AVR dataset (image40.jpg)

### 2.4 Reducing Unclassified Vessels

In order to reduce the number of unclassified vessels the colour features of unclassified vessels were compared quadrant pair wise with that of classified arterioles and venules. Let each image have n vessels, i.e. $S = V_{1k}, V_{2k}, \ldots, V_{nk}$, where $V_{ik}$ is $i$th vessel in $k$th quadrant. Each vessel is represented by a set of m centerline pixels: $V_{ik} = p_{1k}, p_{2k}, \ldots, p_{mk}$. On each centerline pixel of unclassified and classified vessels, MR and MG were extracted because MR and MG are dominating features to distinguish arterioles and venules [9]. The average of MR and MG from all centerline pixels of each of the vessel $V_{ik}$ in all four quadrants ($k = 1, \ldots, 4$) were calculated separately for unclassified vessels ($u$), arteries ($a$) and veins ($v$). This gives [MGu MRu], [MGa MRa] and [MGv MRv]. Then, [MGu MRu] was compared with [MGa MRa] and [MGv MRv] quadrant pair wise in order to assign a label to the unlabeled vessel. For instance, first [MGu MRu] was compared with [MGa MRa] and [MGv MRv] in quadrant pair (I, II). If the absolute difference between [MGu, MRu] and [MGa, MRa], respectively is less than the absolute difference with [MGv, MRv], a soft label of *arteriole* was assigned. Alternatively if the absolute difference between [MGu, MRu] and [MGv, MRv] respectively is less than the absolute difference with [MGa, MRa], a soft label of *venule* was assigned. Likewise the color features of unclassified vessels were compared with that of classified arterioles and venules in remaining quadrant pairs (II, III), (III, IV) and (IV, I) assigning one more soft labels. Finally, each unclassified vessel was assigned a deciding label based on the polling of soft labels. If the vote was tied the vessel remained unclassified.

## 3 Results

Table 1 summarizes the results (where $C$ is the percentage of correctly classified vessels and $Un$ stands for the unclassified vessels, both w.r.t observer 1). The classification rates obtained with SMIC and GMM-EM were almost the same but the percentage of unclassified vessels with GMM-EM is higher compared to SMIC in zone B. Results were slightly higher as compared with observer 2. The classification rates obtained with SMIC in the extended zone were higher and unclassified vessels were lower as compared to GMM-EM.

Table 2 summarizes the classification results when we applied our algorithm to images from INSPIRE-AVR and DRIVE. SMIC outperforms GMM-EM in terms

**Table 1** Classification rate on ORCADES dataset

| Classifier | Zone B C/Un (%) | Extended zone C/Un (%) |
|---|---|---|
| *SMIC* [8] | 90.67/6.48 | 87.66/5.89 |
| *GMM-EM* [13] | 90.45/13.8 | 84.74/13.44 |

**Table 2** Classification rate on publically available dataset in Zone B

| Classifier | INSPIRE-AVR C/Un (%) | DRIVE C/Un (%) |
|---|---|---|
| *SMIC* [8] | 87.6/3.83 | 86.2/7 |
| *GMM-EM* [13] | 83.67/14 | 78.1/38.6 |

**Table 3** Classification rate after applying method of reducing unclassified vessels

| Classifier | Zone B C/Un (%) | Extended zone C/Un (%) |
|---|---|---|
| *SMIC* [8] | 90.62/2.99 | 87.78/2.23 |
| *GMM-EM* [13] | 90.8/4.98 | 85.95/4.47 |

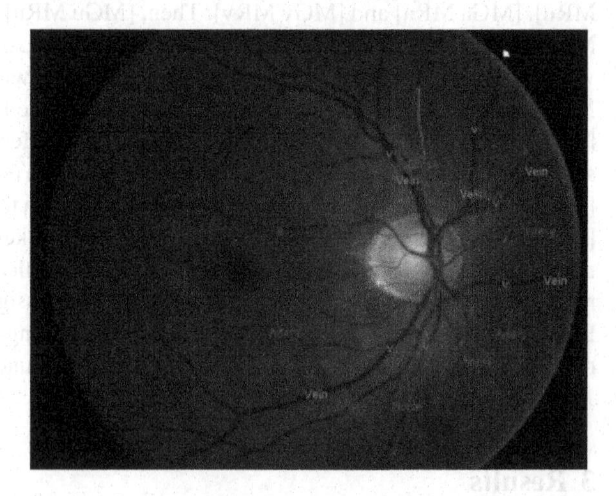

**Fig. 2** Image showing vessel classification: 'v' and 'a' indicates ground truth of 'venule' and 'arterioles' respectively. Vessels marked with 'Vein' and 'Artery' is the classification outcome. 'Noise' are vessel that remain unlabeled. Vessels marked 'Noise' were assigned label 'arteriole' and 'venule' marked in *red* and *blue* respectively after applying our technique for classifying unlabeled vessels

of classification rate and percentage of unclassified vessels. Table 3 shows the result obtained after attempting to reduce the number of unclassified vessels. Figure 2 shows the classification result. There are two vessels which remain unlabeled by the SMIC algorithm and are marked 'Noise' in an image (see Fig. 2). Vessels marked in red and blue in Fig. 2 are 'arteriole' and 'venule' labels respectively assigned to unlabeled 'Noise' vessels after applying our technique for classifying unlabeled vessels.

# 4 Discussion and Conclusion

We have proposed a Squared-loss Mutual Information clustering (SMIC) method to cluster retinal vessels into arterioles and venules for the first time (to the best of our knowledge), using only four color features. The results were very promising in both zone B and in the extended zone. The performance of SMIC was better than that of the GMM-EM unsupervised classifier [13] both in terms of classification rate and unclassified vessels for our dataset as well as publically available datasets. SMIC method clusters the data analytically via kernel eigenvalue decomposition and has advantage of finding the optimal solution. Unlike the previous information-maximization clustering methods, SMIC does not suffer from the problem of local optima. The classification percentage obtained by our system was higher than those reported in [15–18]. We analyzed 70 color fundus images compared to 58, 35 and 15 images analyzed in [16–18], respectively. The resolution of our images was 2048 × 3072 which is higher than 768 × 576, 1300 × 1000 and 800 × 1000 as used in [16–18], respectively. Therefore classification accuracy obtained by our system may differ with a different datasets. Moreover the classification rate is highly dependent on segmentation results (for extracting centerline pixels) and choice of retinal zone, classifier and framework. We validated the performance of our system on INSPIRE-AVR and DRIVE with classification rates of 87.6 and 86.2 %, respectively, which is higher than that obtained with GMM-EM. Our system resulted in a higher classification rate on INSPIRE-AVR when compared to [10] and close to that of [12]. On the DRIVE images the results obtained by our method was close to that of [12] and lower than that of [11]. However, it should be noted that the results in [10–12] are from supervised techniques and our method is unsupervised. Our method of reducing the unclassified vessels reduces the unlabeled vessels and increased the classification rate.

We conclude that our system performance is very promising. Further tests with much larger datasets are needed to declare suitability to support retinal vessel classification in biomarker research.

**Acknowledgments** This work is supported by Leverhulme Trust grant RPG-419 "Discovery of retinal biomarkers for genetics with large cross-linked datasets", and part of the VAMPIRE (Vasculature Assessment and Measurement Platform for Images of the REtina) project led by the University of Dundee and Edinburgh, UK [5]. We thank Dr. Jim Wilson, University of Edinburgh, for making the ORCADES images available.

## References

1. Abràmoff, M.D., et al.: Retinal imaging and image analysis. IEEE Rev. Biomed. Eng. **3**, 169–208 (2010)
2. Wong, T.Y., et al.: Computer-assisted measurement of retinal vessel diameters in the Beaver Dam Eye Study: methodology, correlation between eyes, and effect of refractive errors. American academy of ophthalmology, pp. 90–1183 (2004)
3. Ikram, M.K., et al.: Retinal vessel diameters and risk of hypertension: the rotterdam study, hypertension. J. Am. Hear. Assoc. **47**(2), 94–189 (2006)
4. Leung, H., et al.: Relationships between age, blood pressure, and retinal vessel diameters in an older population. Investig. Ophthalmol Vis. Sci. **44**(7), 2900–2904 (2003)
5. Perez-Rovira, A., et al.: Vampire: vessel assessment and measurement platform for images of the retina. In: Proceedings of the IEEE Engineering in Medicine and Biology Society, pp. 3391–3394 (2011)
6. Li, H., Hsu, W., Lee, M., Wong, T.Y.: Automatic grading of retinal vessel calibre. IEEE Trans. Bio-med. Eng. **52**(7), 5–1352 (2005)

7. Cheung, C.Y.-L., Hsu, W., et al.: A new method to measure peripheral retinal vascular calibre over an extended area. Microcirculation **17**(7), 495–503 (2010)
8. Sugiyama, M., et al.: Information-Maximization Clustering based on Squared-Loss Mutual Information, pp. 40–20111
9. Jelinek, H.F., et al.: Towards vessel characterization in the vicinity of the optic disc in digital retinal images. In: Proceedings of the Image and vision computing (2005)
10. Niemeijer, M., Xu, X.: Automated measurement of the arteriolar-to-venular width ratio in digital color fundus photographs. IEEE Trans. Med. Imaging **30**(11), 50–1941 (2011)
11. Mirsharif, Q.: Automated characterization of blood vessels as arteries and veins in retinal images. Comput. Med. Imaging Graph. **37**(7–8), 17–607 (2013)
12. Dashtbozorg, B., et al.: An automatic graph-based approach for artery/vein classification in retinal images. IEEE Trans. Image Process. **23**(3), 1073–1083 (2014)
13. Relan, D., et al.: Retinal vessel classification: sorting arteries and veins. In: Conference Proceedings: Annual International Conference of the IEEE Engineering in Medicine and Biology Society, pp. 7396–7399 (2013)
14. Relan, D., et al.: Automatic retinal vessel classification using a least square-support vector machine in VAMPIRE'. In: 35th Annual International Conference of the IEEE EMBS Engineering in Medicine and Biology Society (EMBC), pp. 142–145. Chicago, USA (2014)
15. Saez, M., et al.: Development of an automated system to classify retinal vessels into arteries and veins. Comput. Methods programs Biomed. 1–10 (2012)
16. Vazquez, S.G., et al.: On the automatic computation of the Arterio-Venous Ratio in retinal images: using minimal paths for the Artery/Vein classification. In: International Conference on Digital Image Computing: Techniques and Applications, pp. 599–604 (2010)
17. Grisan, E., Ruggeri., A.: A divide et impera strategy for automatic classification of retinal vessels into arteries and veins, In: Proceedings of the 25th Annual International Conference of the IEEE Engineering in Medicine and Biology Society, pp. 890–893 (2003)
18. Joshi, V.S., et al.: Automated artery-venous classification of retinal blood vessels based on structural mapping method. In: Proceedings of the SPIE, vol. 8315 (2012)
19. Staal, J.J., Abramoff, M.D.: Ridge based vessel segmentation in color images of the retina. IEEE Trans. Med. Imaging **23**, 501–509 (2004)
20. Chrástek, R., et al.: Automated segmentation of the optic nerve head for diagnosis of glaucoma. Med. Image Anal. **9**(4), 297–314 (2005)

# Automated Spam Detection in Short Text Messages

**Gaurav Goswami, Richa Singh and Mayank Vatsa**

**Abstract** Increase in the popularity and reach of short text messages has led to their usage in propagating unsolicited advertising, promotional offers, and other unwarranted material to users. This has led to a high influx of such *spam* messages. In order to protect the interests of the user, several countermeasures have been deployed by telecommunication companies to hinder the volume of such spam. However, some volume of spam messages still manage to avoid these measures and cause varying degree of annoyance to users. In this chapter, an automated spam detection algorithm is proposed to deal with the particular problem of short text message spam. The proposed algorithm performs the two class (spam, ham) classification using stylistic and text features specific to short text messages. The algorithm is evaluated on three databases belonging to diverse demographic settings. Experimental results indicate that the proposed algorithm is highly accurate in detecting spam in short messages and can be utilized by a wide variety of users to reduce the volume of spam messages.

## 1 Introduction

*Spam* is defined as an unsolicited and possibly vexatious piece of cost-effective and fast communication which is generally sent in bulk to fulfill promotional needs. The problem of spam detection and classification is one of the staple problems in the machine learning research community. Due to its low cost and ease-of-use for bulk-broadcast of spam messages, e-mail is one of the first communication media to get severely affected by spam. The proliferation of e-mail spam instigated the initial interest in this area of research and a considerable amount of literature deals

G. Goswami (✉) · R. Singh · M. Vatsa
Indraprastha Institute of Information Technology, Delhi, India
e-mail: gauravgs@iiitd.ac.in

R. Singh
e-mail: rsingh@iiitd.ac.in

M. Vatsa
e-mail: mayank@iiitd.ac.in



R. Singh et al. (eds.), *Machine Intelligence and Signal Processing*,
Advances in Intelligent Systems and Computing 390,
DOI 10.1007/978-81-322-2625-3_8

with the problem of identifying and filtering spam e-mail messages. In recent past, spammers have found an even more accessible and increasingly popular medium of communication, Short Message Service (SMS). The popularity of SMS as a mode of communication has increased over the years and is now widely used for quick communication via text messages. With 20–30 % spam in SMS traffic, it is a major problem in several Middle Eastern and Asian countries including Europe.

SMS spam can sometimes be more annoying than e-mail spam due to its realtime and urgent nature. While e-mails are read leisurely and provide a subject line that helps users identify spam very quickly and ignore such messages, the same does not hold true for SMS spam. Users tend to disrupt their current task in order to read a message as it arrives and makes the notification sound. Around 95 % text messages received are opened and read within minutes. If there are multiple such instances in short periods of time, it can become a nuisance. Moreover, sometimes such messages arrive at odd hours and cause further annoyances. If an automated algorithm can be built which can detect incoming spam messages accurately and notify the user only for non-spam messages, then the user can be relieved of the problem of SMS spam. The smartphone market has been steadily moving towards providing many conveniences to the users with features such as gesture-based control and automatic turning off of screen becoming sufficiently significant to become the unique selling points of devices. Protection against spam is another in the list of such conveniences that can be attractive to a user and therefore, there is high motivation to design such an algorithm.

There exist several challenges in designing an algorithm to detect short message spam when compared to e-mail spam. First, these messages are short and mostly below 160 characters. Many messages can be far below this limit, less than even hundred characters. Therefore, the lack of content available for analysis and feature extraction is a challenge on its own. Second, short messages are written using shorthand versions of long words which may vary from user to user. For example, users might abbreviate the word 'communication' as *communicatn*, *commn*, or *comnctn*. There can be many different variants of a single word that can be found in such messages. This heterogeneity makes it difficult to trace these abbreviations back to the original word and introduces problems if the words are used as features. Third, in a multi-lingual country like India, a single SMS can be comprised of words belonging to different languages written in the English script. For example, a typical SMS can begin with a 'Hi' in English, and then the rest of the content can be written in Hindi ('*kya kar rha hai?*' i.e., what are you doing?). Also, when adapting a language based on a script other than English to its alphabet, various forms of the same word are introduced, e.g., the Hindi word for 'yes' can be written as *haa, haan, han, ha,* etc. depending on the user's judgment. This again causes similar problems when different users use different abbreviations. These situations create challenges for a tokenization-based approach. A tokenization-based approach operates by extracting the words in the SMS as tokens. The frequency of particular tokens is characterized in order to classify messages as not-spam or spam. In addition, low amount of tokens can lead to a decision influenced by just one or two tokens. This can lead to a decision below the confidence threshold of a classifier, and therefore the prediction cannot be utilized. Besides these problems, there is a fundamental issue with the availability of labeled SMS data for training the classifiers.

In this research, a collection of stylistic features are utilized for SMS spam classification. These features are not influenced by the content itself but by the style and composition of the message content. The proposed spam detection algorithm utilizes a combination of classification models based on both stylistic and text features. Classification results using these features are presented on three demographically diverse datasets. The experimental results indicate that the proposed algorithm accurately identifies both ham and spam messages.

## 2 Literature Review

Most of the existing research in spam detection has focused in the context of e-mail messages. However, there have been some publications in the domain of short text messages as well. Cormack et al. [3] deal with three kinds of short messages: blog comments, e-mail summaries, and mobile SMS communications. They have used various classification models with character level features such as bigrams, trigrams, and sparse word bigrams for classification. They have also mentioned the use of a few stylistic features but have not provided the entire details of which stylistic features are chosen in the experiments. The dataset size used for mobile SMS communication has 1002 legitimate messages and 322 spam messages. Xiang et al. [16], Healy et al. [4], and Hidalgo et al. [6] have adapted automatic text classification to the problem of SMS spam filtering. They conclude that Support Vector Machines (SVMs) perform the best among the evaluated classifiers. Longzhen et al. [10] propose a k-nearest neighbor algorithm ($k$-NN)-based multi-level filtering approach that is evaluated on 550 spam and 200 ham messages. Liu et al. [9] propose a simple index ensemble model based on the lexical analysis. Junaid et al. [8] discuss the application of evolutionary classifiers for SMS spam filtering and compare supervised approaches with evolutionary approaches but find that the performance of supervised approaches is comparable and faster. Almeida et al. [1] compare a variety of supervised approaches and also report that SVM performs the best on the spam classification problem.

Besides the server-based solutions, a few client-based approaches have also been proposed. Deng et al. [5] propose a Naive Bayes-based filtering system that utilizes user feedback to improve over time with central server-based retraining. Among recent works, Rafique et al. [14] utilize a byte-level representation of the messages to train first-order Hidden Markov Models (HMMs) to estimate the probabilities of byte sequences in spam and ham messages. Yadav et al. [17] propose a total of 20 stylistic features for representing short messages in order to reduce the computation time for classification so that the program can run on the mobile devices as a client-side application. The top 10 stylistic features sorted by their F scores have been reported in the paper. They achieve 93 % ham (not-spam) and 86 % spam accuracy while using the stylistic features with SVM learning, and 97 % and 72.5 % when

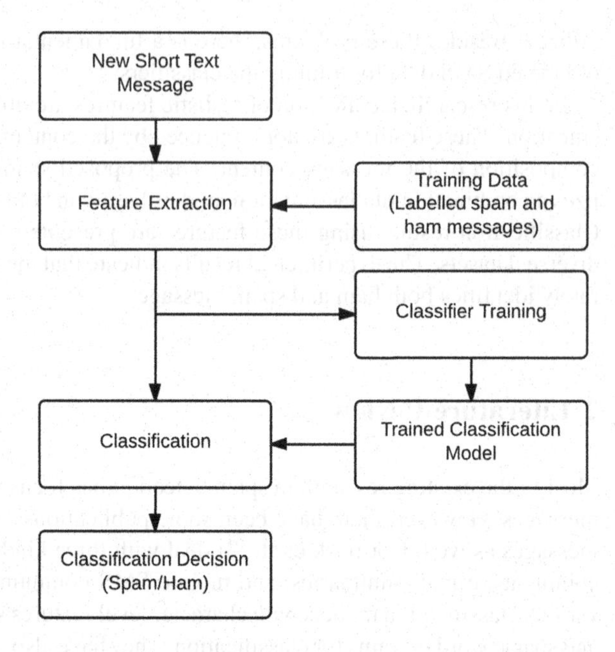


**Fig. 1** The steps involved in the SMS Spam Detection Algorithm

using them with Bayesian learning. The size of the dataset used in this work contains 2195 ham and 2123 spam messages (Fig. 1).

A few solutions also operate at cellular operator level. Recent research has investigated spam filtering in this manner and evaluated the approaches on the basis of datasets provided by cellular companies. Xu et al. [13] propose a number of statistics such as message length, response ratio, and number of messages, and temporal features such as messages per day, messages during a particular hour of day, and average number of recipients for a week to characterize cellular service users as spammers/legitimate users. Jiang et al. [7] propose a method based on characterizing users on the basis of a multitude of features including text analysis of sent messages, usage patterns, and device characteristics. Murynets et al. [11] analyze data provided by a tier-1 cellular operator in the United States with features similar to those proposed in [13].

## 3 Proposed Spam Detection Algorithm

Figure 3 presents an overview of the proposed algorithm. During training, features are extracted from the labeled dataset and a classification model is created. During testing, each new short text message is classified on the basis of features extracted from it and the trained classification model. The short messages are classified into two categories: spam and ham (not-spam). The following sections describe the various features utilized by the proposed approach and provide further insight into its functionality.

## 3.1 Preprocessing

The content of the text messages is used without any preprocessing for feature extraction. The scope of preprocessing in a short text message is very limited as the content itself is short: 160 characters in a single message. Most spam messages are tailored to be one screen page long only as messages are charged on a per page basis. Sometimes, ham messages can be less than even half of this character limit. Since most of the terms are going to be 'common,' we chose not to use stop word removal. At the best, stop word removal serves the purpose of trimming down the possible features, thereby reducing the dimensionality of the feature space and also reducing the size of the vocabulary. We choose to ignore this storage advantage as it does not have a major practical impact on the usability of the system. Another popular technique of preprocessing text data before classification is stemming, whereby the words are reduced to their root forms (e.g. clarification, clarifying, clarified → clarif). Since the text messages are written using short forms of the words, these may not be consistent across senders and long words are infrequent. Therefore, we choose to avoid stemming as well.

## 3.2 Stylistic Features

As discussed previously, text features are prone to many challenges when considering the problem of short text messages. On the other hand, stylistic features are based on the prior knowledge that personal messages labeled as ham are written differently when compared to advertising (spam) messages. These features exploit the basic stylistic differences in order to provide features for classification. Since these features are not dependent directly on the text of the message but on overall statistics of the content, these tend to be robust towards language variations. Moreover, these features are faster to extract, easier to store, and also lead to faster classification due to lesser dimensionality of feature vectors. In the proposed algorithm, the following stylistic features are utilized:

1. Caps Count (count of uppercase characters): Spam messages generally tend to have high uppercase content in order to highlight terms such as "LIMITED TIME OFFER," "HURRY," and "DISCOUNT," whereas in a general ham message, the user tends to type in all lowercase characters (default mode in majority of phones).
2. Caps Only Word Count (count of words that are composed of uppercase characters only): Some phones automatically capitalize the first character of the first word of a sentence. Therefore, there is bound to be some uppercase content in the ham messages as well. This feature is designed to account for this situation.
3. Digit Count (count of digits in the message body): Spam messages include a phone number to call, some figures (such as prices and timings). Therefore, it is intuitive to check for the presence of digits in the message.

4. Digit Only Word Count (count of words entirely composed of digits): This feature detects the presence of a full phone number, price, etc. Sometimes users use numbers in abbreviations of common words (before -¿ b4, etc.). This feature takes care of such scenarios in case of ham messages.
5. Emoticon Count (count of emoticons in the message): Spam messages rarely have emoticons such as :), :(, and :-). In fact, in the datasets used for this work, no spam message contained an emoticon. Therefore, this feature is a very strong indicator of a message being spam or not.
6. Special Character Count (count of special characters in the message): This set of special characters is based on contextual knowledge about spam messages. Characters found in URLs (/,:), in percentages (%), and in prices (Rs., $, etc.) are included in special character set. Ham messages tend to have a really small amount of special characters, if any.
7. Single Character Word Count (count of words comprising a single character): These words are mostly present in ham messages (are-*¿r, you-¿u, we-¿v*, etc.). Therefore, this feature can also be used in the stylistic attribute set.

Some of the proposed features are in previous works such as [17]. However, there are two major differences in the proposed set:

1. The proposed feature set is a very concise set of only seven features.
2. These features are also normalized by dividing the counts by the total number of words in the message. Another scheme that is also evaluated is to normalize the character counts by the number of characters and the word counts by the number of words in the message. However, during experimentation, the former strategy is found to perform better.

## *3.3 Text Features*

In addition to stylistic features, text features are also included in the proposed algorithm. Text features have been explored extensively for e-mail spam detection. Besides providing valuable data for performance comparison, text features provide an additional source of feature information. Text features are explored in two ways: terms as features and character level n-grams. Actual terms are used as features which may provide some discriminability. On the other hand, character level n-grams are chosen since they are comparatively more robust to the challenges discussed previously compared to directly using terms. Even though different abbreviations of a word are different, these still may have some n-grams in common. For example, attention and attn both share the two grams 'at' and 'tt.'

### 3.4 Classification

In order to utilize the stylistic features for classification, first the values of the features are computed. Then, these feature values are arranged in a pre-determined manner in order to create a feature vector for the message. These feature vectors can be generated for both existing and new messages, and hence classification can be performed using any vector classification algorithm. In case of text features, each text message can be converted to a feature vector using the popular tf-idf approach. Once these feature vectors are generated, two Support Vector Machines (SVMs) [2] are trained on the stylistic features and the text features. The decisions of these two classifiers are combined such that a test message is considered spam if and only if both the classifiers classify it as spam. Otherwise, the message is considered to be ham. This methodology ensures that the approach prioritizes correct classification of ham messages.

## 4 Database and Experimental Protocol

Three datasets are used in this research. Dataset 1 is the dataset used in [17] with some mislabeled data removed. This dataset consists of 2,024 ham messages and 2,113 spam messages. It has words in both English and Hindi (written using English alphabet) as are generally used in the Indian context. Dataset 2 is the SMS Spam Collection v.1 [15] which comprises 747 spam messages and 4,827 ham messages. This dataset comprises only English words. Dataset 3 is prepared by the authors of [12] and contains 1450 tagged messages (730 ham and 721 spam) and a test set of 700 messages.

In order to highlight the differences and similarities between the message content in these different demographic settings, we utilize word clouds. The word cloud is a method to visualize text data in which the most frequently occurring words are presented in an image format, as shown in Fig. 2. The size of each word is proportional to its relative frequency in the text. It can be seen that the most prominent words in the ham word clouds are common words that are used in daily communication. Besides, the word cloud from dataset 1 has a large number of non-English words written using the English alphabet, whereas datasets 2 and 3 have only English words. In contrast to the ham word clouds, these clouds contain more common words such as 'call' and 'free.' Compared to the word distribution of the ham messages which are quite different, there are noticeable similarities in case of spam messages even in different demographic settings. There are words related to limited time offers, free prizes, and some brand names. Some of these words are also common in both word clouds. In our experiments, 40 % of each dataset is utilized for training and the remaining 60 % is used for testing. Since it is important to not misclassify ham messages as spam, the parameters of all algorithms are tuned to prioritize ham accuracy. The reported results for each approach report the spam accuracy achieved at the respective ham accuracy.

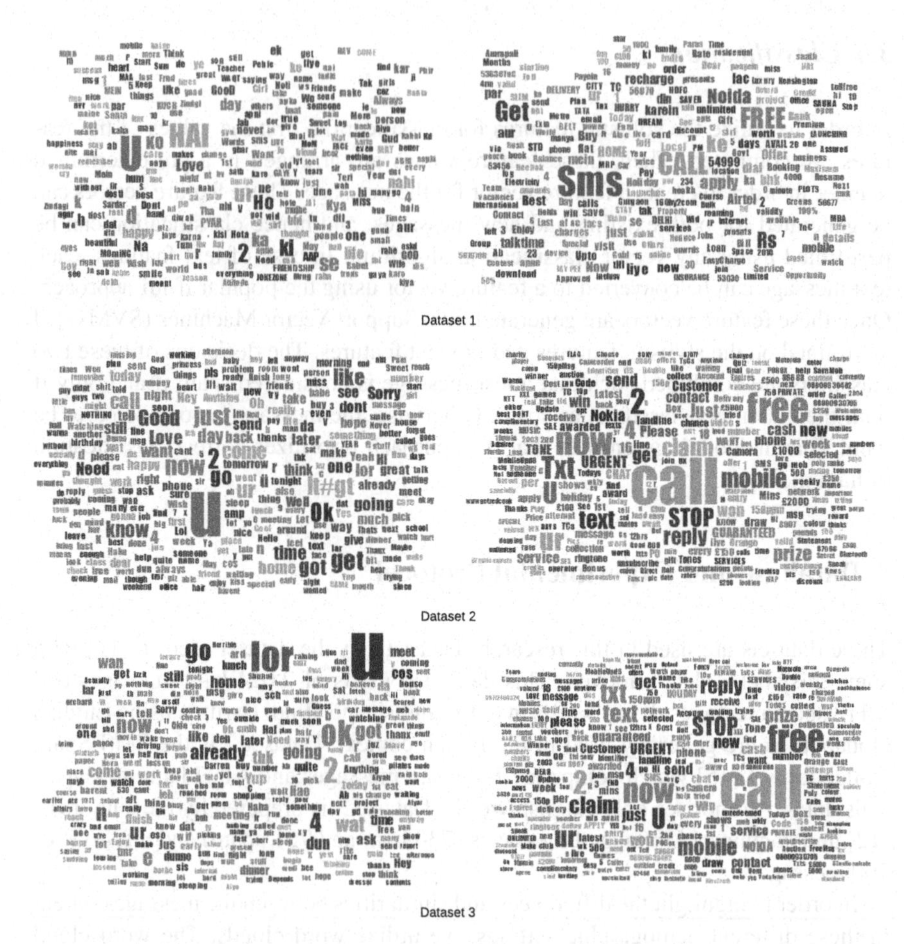


**Fig. 2** Word clouds for all three datasets. The *right column* contains the spam word clouds and the *left column* contains the ham word clouds

# 5 Results

The mean values of the stylistic features for the datasets are reported in Table 1. These values have been calculated using all the data points in the individual datasets. The feature values are normalized by either the number of words in the message or the message length (number of characters in the message). The normalization varies from feature to feature and the exact type has been indicated in the table. The results in this table illustrate the distribution of the values of these features for the two classes (spam and ham) across the three datasets. This distribution indicates that the values of these features vary according to the class and therefore can be useful for classification. The relatively low variance of the values coupled together with high difference in the mean values for these features for the other class ensure high inter-

**Table 1** Mean values for stylistic features

| Feature | Normalized by | Ham | | | Spam | | |
|---|---|---|---|---|---|---|---|
| | | Dataset 1 | Dataset 2 | Dataset 3 | Dataset 1 | Dataset 2 | Dataset 3 |
| Caps count | No. of chars | 0.09 ± 0.01 | 0.24 ± 0.25 | 0.16 ± 0.02 | 0.15 ± 0.02 | 0.53 ± 0.18 | 0.56 ± 0.25 |
| Caps only word count | No. of words | 0.05 ± 0.01 | 0.04 ± 0.01 | 0.02 ± 0.01 | 0.11 ± 0.02 | 0.08 ± 0.01 | 0.09 ± 0.01 |
| Digit count | No. of chars | 0.01 ± 0.01 | 0.01 ± 0.00 | 0.04 ± 0.01 | 0.09 ± 0.00 | 0.58 ± 0.14 | 0.57 ± 0.18 |
| Digit only Word Count | No. of words | 0.00 ± 0.00 | 0.00 ± 0.00 | 0.00 ± 0.00 | 0.07 ± 0.02 | 0.07 ± 0.00 | 0.07 ± 0.00 |
| Emoticon count | No. of chars | 0.00 ± 0.00 | 0.01 ± 0.00 | 0.00 ± 0.00 | 0.00 ± 0.00 | 0.00 ± 0.00 | 0.00 ± 0.00 |
| Special character count | No. of words | 0.00 ± 0.00 | 0.00 ± 0.00 | 0.00 ± 0.00 | 0.00 ± 0.00 | 0.02 ± 0.00 | 0.01 ± 0.00 |
| Single character word count | No. of words | 0.08 ± 0.00 | 0.08 ± 0.00 | 0.10 ± 0.01 | 0.05 ± 0.00 | 0.050 ± 0.00 | 0.06 ± 0.00 |

**Table 2** Correlation between all stylistic features

| | Caps count | Caps only word count | Digit count | Digit only word count | Emoticon count | Single char word count | Special char count |
|---|---|---|---|---|---|---|---|
| Caps count | | 0.884 | 0.287 | 0.244 | –0.052 | 0.014 | 0.167 |
| Caps only word count | | | 0.191 | 0.183 | –0.083 | –0.005 | 0.106 |
| Digit count | | | | 0.814 | 0.042 | –0.138 | 0.4007 |
| Digit only word count | | | | | 0.024 | –0.126 | 0.353 |
| Emoticon count | | | | | | 0.143 | –0.049 |
| Single char word count | | | | | | | 0.0019 |

class variation and low intra-class variation which is a prerequisite for discriminative features.

Table 2 provides the correlation between the features. In order to compute the correlation values, the feature values from the datasets are combined. A correlation value close to 1 indicates a strong positive relationship, whereas a value close to $-1$ indicates a strong negative relationship between the two concerned features. A value close to 0 indicates that the features are independent of each other. Independent features are important for the purposes of classification since they provide more information to the classifier. As expected, the only highly correlated feature pairs are (1) Caps Count and Caps Only Word Count and (2) Digit Count and Digit Only Word Count, since these are supposed to be directly related to each other. The other feature pairs have low correlation and therefore the overall feature set is suited for classification. Moreover, to visualize the separability of the features in feature space, we have also analyzed the scatter plots. The best three feature pairs (lowest correlation) are selected to generate scatter plots in order to visualize the effectiveness of the stylistic features. Figures 3, 4, and 5 illustrate the scatter plots of the ham and spam data against the corresponding feature sets: (Special Character Word

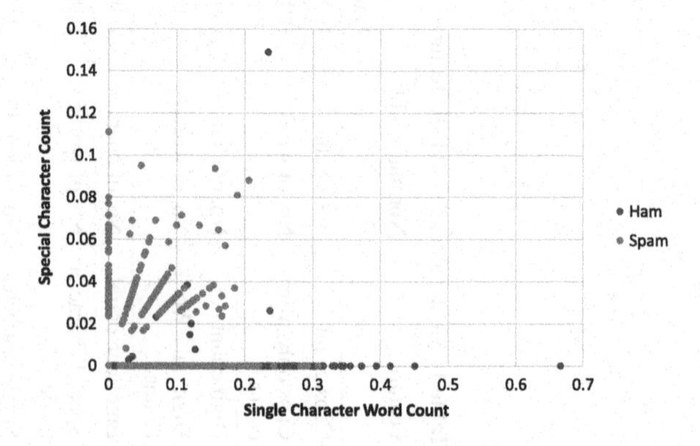


**Fig. 3** Scatter of ham and spam data points against Special Character Count and Single Character Word Count

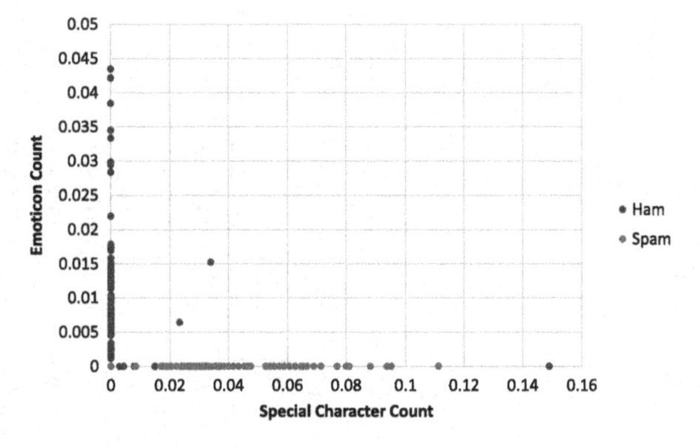


**Fig. 4** Scatter of ham and spam data points against Special Character Count and Emoticon Count

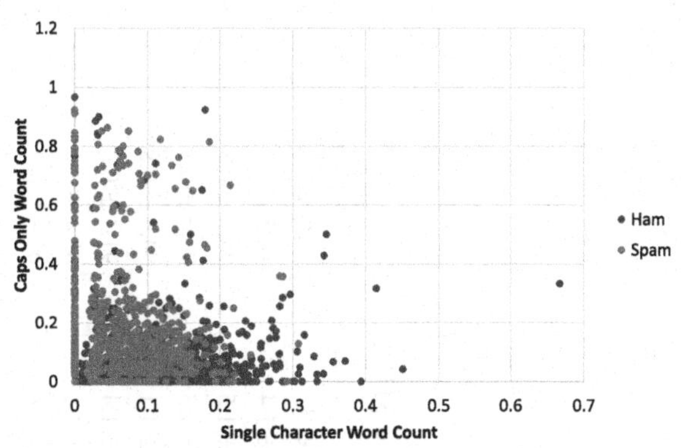


**Fig. 5** Scatter of ham and spam data points against Single Character Word Count and Caps Only Word Count

Count vs. Single Character Word Count), (Special Character Count vs. Emoticon Count), and (Single Character Word Count vs. Caps Only Word Count). The data points are highly separable in the first two cases and moderately separable in the third case. On the basis of these scatter plots, it can be inferred that these features in conjunction with a classifier can be used to distinguish between ham and spam messages. Moreover, it is evident from Figs. 3 and 4 that when the values of the counts are non-zero, it is easier to distinguish between ham and spam. However, it is challenging to distinguish when one or more of these counts are zero. Therefore, all the features are utilized in conjunction to perform the classification so that even if some counts are zero (provide no information) the other features can be used to perform classification. Furthermore, these plots also indicate that ham and spam are not linearly separable, and hence a non-linear decision boundary might be required to distinguish effectively between the two.

Table 3 summarizes the results, providing mean and standard deviation for ham and spam accuracies over 10 times random subsampling-based cross validation. The results indicate that the proposed stylistic features yield high accuracy in detecting SMS spam messages. In addition to the proposed algorithm, the performances of other classifiers are also evaluated. The proposed approach, utilizing a combination of stylistic- and text-based features, achieves the best performance across all the

**Table 3** Classification results on Dataset 1 [17], Dataset 2 [15], and Dataset 3 [12]

| Algorithm | HAM accuracy | | | SPAM accuracy | | |
|---|---|---|---|---|---|---|
| | Dataset 1 | Dataset 2 | Dataset 3 | Dataset 1 | Dataset 2 | Dataset 3 |
| Bayes (n-grams) | 93.3 ± 1.2 | 46.1 ± 1.3 | 92.1 ± 1.4 | 96.5 ± 0.7 | 99.4 ± 0.4 | 91.2 ± 1.8 |
| Bayes (terms) | 88.6 ± 1.4 | 50.1 ± 1.6 | 90.2 ± 1.5 | 98.3 ± 0.8 | 99.7 ± 0.9 | 91.5 ± 2.1 |
| SVM (n-grams) | 99.2 ± 0.4 | 99.5 ± 0.3 | 92.1 ± 1.3 | 96.0 ± 0.4 | 92.5 ± 0.1 | 90.2 ± 1.2 |
| SVM (terms) | 97.2 ± 1.1 | 99.4 ± 0.6 | 90.1 ± 1.1 | 94.0 ± 0.8 | 90.6 ± 0.1 | 90.7 ± 0.7 |
| SVM (stylistic) | 95.2 ± 0.2 | 96.1 ± 0.5 | 91.4 ± 1.4 | 91.3 ± 0.3 | 89.1 ± 0.7 | 89.6 ± 1.4 |
| Rocchio (stylistic) | 71.5 ± 1.6 | 89.0 ± 1.1 | 84.3 ± 2.1 | 51.4 ± 1.3 | 87.0 ± 1.1 | 80.2 ± 2.4 |
| RDF (stylistic) | 94.0 ± 0.4 | 92.1 ± 0.7 | 92.2 ± 1.6 | 91.5 ± 0.2 | 95.0 ± 0.5 | 90.3 ± 1.5 |
| Proposed | 99.4 ± 0.2 | 99.3 ± 0.5 | 98.7 ± 0.0 | 98.3 ± 0.3 | 96.5 ± 0.1 | 99.0 ± 0.0 |

three datasets. Due to the lack of standard training testing partitions (protocol) and unavailability of their implementations (or open source), it is difficult to directly compare accuracies with previous results on these databases. However, [12] provides a standard test set of around 300 spam and 400 ham messages in which the proposed algorithm is as good as the current state-of-the-art. Computationally, the proposed algorithm is very fast and under C# programming environment on an Intel Core i7 machine @ 2.1 GHz, it requires around 10 ms to classify one SMS message.

## 6 Conclusion and Future Work

This paper presents a computationally efficient SMS spam detection algorithm which utilizes stylistic and text features in conjunction with SVM classifier. This research shows that the spam detection problem can be addressed reliably even without using direct content clues such as n-grams and words. The experiments suggest that the methods which characterize the way in which a message is written can also be used for discriminating spam from ham messages. The proposed algorithm yields state-of-the-art results on three SMS spam–ham databases. The future research in this area might focus on expanding and further strengthening the language-independent stylistic feature set.

**Acknowledgments** The authors would like to thank the authors of [17] for providing the SMS dataset.

## References

1. Almeida, T.A., Hidalgo, J.M.G., Yamakami, A.: Contributions to the study of sms spam filtering: new collection and results. In: ACM Symposium on Document Engineering, pp. 259–262 (2011)
2. Chang, C.C., Lin, C.J.: LIBSVM: A library for support vector machines. ACM Trans. Intell. Syst. Technol. **2**(3) (2011)
3. Cormack, G.V., Gómez Hidalgo, J.M., Sánz, E.P.: Spam filtering for short messages. In: Conference on Information and Knowledge Management, pp. 313–320 (2007)
4. Delany, S.J., Zamolotskikh, A.: An assessment of case base reasoning for short text message classification (2004)
5. Deng, W.W., Peng, H.: Research on a naive bayesian based short message filtering system. In: International Conference on Machine learning and cybernetics, pp. 1233–1237 (2006)
6. Gómez Hidalgo, J.M., Bringas, G.C., Sánz, E.P., García, F.C.: Content based sms spam filtering. In: ACM Symposium on Document Engineering, pp. 107–114 (2006)
7. Jiang, N., Jin, Y., Skudlark, A., Zhang, Z.L.: Understanding sms spam in a large cellular network: characteristics, strategies and defenses. Res. Attacks Intrusions Defenses **8145**, 328–347 (2013)
8. Junaid, M.B., Farooq, M.: Using evolutionary learning classifiers to do mobilespam (sms) filtering. In: Genetic and Evolutionary Computation Conference, pp. 1795–1802 (2011)
9. Liu, W., Wang, T.: Index-based online text classification for sms spam filtering. J. Comput. **5**(6), 844–851 (2010)

10. Longzhen, D., An, L., Longjun, H.: A new spam short message classification. Int. Workshop Educ. Technol. Comput. Sci. **2**, 168–171 (2009)
11. Murynets, I., Piqueras Jover, R.: Crime scene investigation: sms spam data analysis. In: ACM Conference on Internet Measurement Conference, pp. 441–452 (2012)
12. Narayan, A., Saxena, P.: The curse of 140 characters: evaluating the efficacy of sms spam detection on android. In: Third ACM Workshop on Security and Privacy in Smartphones and Mobile Devices, pp. 33–42 (2013)
13. Qian, X., Evan, W.X., Yang, Q.: Sms spam detection using non-content features (2012)
14. Rafique, M.Z., Farooq, M.: Sms spam detection by operating on byte-level distributions using hidden markov models (hmms). In: Virus Bulletin International Conference (2010)
15. The sms spam collection v1. http://www.dt.fee.unicamp.br/tiago/smsspamcollection/
16. Xiang, Y., Chowdhury, M., Ali, S.: Filtering mobile spam by support vector machine. In: Conference on Computer Sciences, Software Engineering, Information Technology, E-Business and Applications, pp. 1–4 (2004)
17. Yadav, K., Kumaraguru, P., Goyal, A., Gupta, A., Naik, V.: Smsassassin: crowdsourcing driven mobile-based system for sms spam filtering. In: Mobile Computing Systems and Applications, HotMobile, pp. 1–6 (2011)

# Greedy Algorithms for Non-linear Sparse Recovery

**Kavya Gupta and Angshul Majumdar**

**Abstract** In this work we propose two greedy algorithms to solve the underdetermined nonlinear sparse recovery problem. There are hardly any algorithms to solve such problems. Our algorithm is based on the OMP (Orthogonal Matching Pursuit) and the CoSaMP (Compressive Sampling Matching Pursuit), two popular techniques for solving the linear problem. We empirically test the success rates of our algorithms on exponential and logarithmic functions. The success rates follow a pattern similar to the one we observe for linear recovery problems.

**Keywords** Compressed sensing · EEG · WBAN

## 1 Introduction

In the last decade, there has been a lot of interest in solving linear inverse problems where the solution is known to be sparse (1), especially when the problem is underdetermined.

$$y_{m\times 1} = A_{m\times n}x_{n\times 1} \tag{1}$$

It is assumed that the solution x is k-sparse, i.e., has only k nonzero elements.

If one has an oracle, and thereby knows the positions of the nonzero indices, the problem becomes trivial. One only needs to solve the linear problem on the reduced set, i.e.,

$$y = A_{\Omega}x_{\Omega} \tag{2}$$

K. Gupta · A. Majumdar (✉)
Indraprastha Institute of Information Technology,
New Delhi 110020, India
e-mail: angshul@iiitd.ac.in
URL: http://www.iiitd.ac.in

K. Gupta
e-mail: kavya1482@iiitd.ac.in



R. Singh et al. (eds.), *Machine Intelligence and Signal Processing*,
Advances in Intelligent Systems and Computing 390,
DOI 10.1007/978-81-322-2625-3_9

Assuming $\Omega$ as the oracle set of indices, solving (2) is trivial via least squares. One thus gets the nonzero values in x; the rest $n - \Omega$, positions in x are filled by zeroes.

Unfortunately for all practical problems, the positions of the nonzero indices are not known *a priori*. In such a case, the only way to solve (1) is via exhaustive search, i.e., evaluate all ${}^{n}C_{k}$ possible combinations and solve (2) for each of these until one finds the solution. As can be seen, this is a combinatorial problem and therefore cannot be solved in practice.

In his seminal work, Donoho [1] showed that sparse solutions are necessarily unique, i.e., there cannot be two k-sparse solutions or there cannot be two solutions that are k-sparse and $k + 1$ sparse. Therefore instead of trying to find a k-sparse solution, one may as well find the sparsest solution to (1). This is expressed as:

$$\min_{x} \| x \|_0 \ s.t. y = Ax \tag{3}$$

This is still a combinatorial problem.

There are two approaches to solve the sparse recovery problem. The first approach is to relax the NP hard $l_0$-norm by its nearest convex surrogate the $l_1$-norm:

$$\min_{x} \| x \|_1 \ s.t. y = Ax \tag{4}$$

Such $l_1$-norm minimization problems can be solved using linear programming. The $l_1$-norm minimization is very popular in applied Compressed Sensing (CS).

There is another approach to solve the sparse recovery problem (1) greedy algorithms. The greedy techniques are based on iteratively populating the index set ($\Omega$) via some heuristic and solving for the corresponding nonzero values. Orthogonal Matching Pursuit (OMP) [2] and CoSaMP [3] are the two most popular greedy recovery algorithms.

In this work, our interest is in solving nonlinear sparse recovery problems of the form:

$$y = f(x), \quad f : \mathbb{R}^n \to \mathbb{R}^m \tag{5}$$

In a previous study [4], a nonlinear $l_1$-norm minimization problem was proposed to solve (5). The problem with such an approach is its slow convergence. Greedy algorithms have always triumphed over $l_1$-norm minimization when speed was concerned, e.g., OMP converges in exactly k iterations. We will discuss more about convergence issues of $l_1$-norm minimization in a later section.

To overcome the limitations of speed, we propose greedy algorithms based on the modifications to OMP and CoSaMP algorithms to solve the nonlinear sparse recovery problem (5). But before going into our proposed technique, we will discuss a practical motivation of such nonlinear sparse recovery problems in the next section. Prior art on nonlinear sparse recovery is discussed in Sect. 3. The proposed techniques are described in Sect. 4. The experimental results are shown in Sect. 5. Finally, the conclusions of our work are discussed in Sect. 6.

## 2 A Practical Motivation

There are several areas that benefit from nonlinear sparse recovery problem, but owing to limitations in space, we will discuss a pertinent from Magnetic Resonance Imaging (MRI). Image intensity for a T1 weighted image is given as follows:

$$x = \rho(1 - exp^{-TR/t}) \tag{6}$$

Here $\rho$ is the proton density, TR is the repetition time, and t is the T1 value for the tissue. In quantitative MRI, the interest is in estimating the proton density and the T1 values from x.

The proton density map is estimated by applying a large repetition time, so as to eliminate the effects of T1 weighting, so that x = $\rho$ under such a circumstance. Once the proton density is known, the interest is in estimating the T1 values, i.e., estimating t.

The T1 map for the entire field of view, is sparse in a transform domain-like wavelet (W), i.e., t = Wx is sparse. Incorporating this sparsity into (6) leads to:

$$x = \rho(1 - exp^{-TR/W^T\alpha}) \tag{7}$$

The sparse coefficient of the T1 map can be solved by:

$$\min_{\alpha} \| \alpha \|_1 \; s.t. x = \rho(1 - exp^{-TR/W^T\alpha}) \tag{8}$$

This is a typical candidate for nonlinear sparse recovery. There are other examples of similar nature. The interested reader can peruse [4].

## 3 Literature Review

Literature on nonlinear sparse recovery is itself sparse. There are two studies [5, 6] of theoretical nature that explores the conditions under which such recovery is possible. They do not propose practical algorithms for solving such problems. To the best of our knowledge, there is only a single work [3] from 2008 that claimed to propose a greedy algorithm to solve the nonlinear sparse recovery problem. However, in essence they proposed an algorithm to recover a sparse solution to a linear problem where the cost function is not Euclidean.

The only algorithmic work on solving nonlinear sparse recovery problems is [4]. There the authors proposed an $l_1$-norm minimization approach; the recovery was posed as:

$$\min_{x} \| x \|_1 \; s.t. \| y - f(x) \|_2^2 \leq \epsilon \tag{9}$$

Solving (9) yields the solution for (5), for a very small value of epsilon. The algorithm in [4] is based on the Iterative Soft thresholding Algorithm. It is a two-step algorithm. The first step is gradient descent that partially solves the nonlinear least squares problem:

$$b = x_{k-1} - \sigma \nabla_x \parallel y - f(x) \parallel_2^2 |_{x=x_{k-1}}$$

where sigma is the step size. The second step is to threshold b in order to project it on the $l_1$-ball, thereby yielding a sparse solution:

$$x_k = soft\ Th(b, \tau)$$

The problem with such an optimization-based approach is that it is not nonparametric. For the linear case, it is easy to find the step size and the threshold $\tau$; but for nonlinear cases it is not. In [4], the step size was determined by the inverse of the Lipchitz bound of the cost function. But it is a very pessimistic (small) step size and the algorithm has very slow (linear) convergence rate.

We find two issues with the previous optimization-based approach for nonlinear sparse recovery,

1. The method is parametric
2. Analytical estimates of the parameters lead to very slow convergence.

To overcome these limitations we propose greedy algorithms to solve the nonlinear sparse recovery problem. Unlike the previous study [7], our algorithm will be able to solve nonlinear problems.

# 4 Proposed Greedy Algorithm

Before discussing the nonlinear extensions, let us discuss the greedy algorithms for linear sparse recovery. The most commonly used algorithm for the purpose is Orthogonal Matching Pursuit (OMP). The algorithm is as follows:

OMP Algorithm:-

| Initialize: Index set - $\Omega$=$\emptyset$,x=0<br>Repeat for k iterations<br>Compute correlation - c=$abs(A^T(y - Ax))$<br>Select the index - i=$\underset{i}{\text{argmax}} c_i$<br>Update support - $\Omega$=$\Omega \cup i$<br>Estimate nonzero values - $x_\Omega$=$\min_x \parallel y - A_\Omega x \parallel$<br>Finally, impute other indices in x with zeroes. |
|---|

There are two issues with the OMP algorithm. First, it only selects one index in every iteration. This is slow, there are simple techniques to accelerate OMP, notably

Stagewise OMP [8]. However, slow selection is not the most pressing problem. The bigger problem is that an index once selected remains there; OMP does not have the scope of pruning an incorrectly selected index at a later iteration. To overcome both these issues, the CoSaMP algorithm was proposed.

CoSaMP Algorithm:-

| Initialize: Index set - $\Omega=\emptyset$,x=0<br>Repeat until convergence<br>Compute correlation - c=$abs(A^T(y-Ax))$<br>Select the top 2 K support from c - $\omega$<br>Update support - $\Omega=\Omega \cup \omega$<br>Perform least squares estimate- $b_\Omega=\min_x \| y - A_\Omega x \|$<br>Update x by pruning b and keeping top K values.<br>Finally, impute other indices in x with zeroes. |
|---|

CoSaMP allows for selection of multiple indices in every iteration (top 2k) and it allows for pruning (by keeping top k).

Let us look at the computing of correlation for both the algorithms. It is computed as $(A^T(y-Ax))$; this is the negative gradient of the cost function $\| y - Ax \|_2^2$. In OMP, we want to solve the linear sparse recovery problem, i.e.,

$$x : y = Ax \;\; s.t. \| x \|_0 = k \tag{10}$$

The equality (y = Ax) holds only at convergence, for all other iterations (in OMP and CoSaMP) one finds x by minimizing the Euclidean distance $\| y - Ax \|_2^2$.

Our idea for nonlinear sparse recovery is based on the same approach. For our problem, the Euclidean cost function is $\| y - f(x) \|_2^2$. We would like to solve the nonlinear sparse recovery problem:

$$x : y = f(x) \;\; s.t. \| x \|_0 = k \tag{11}$$

Therefore, our correlation should be defined as

$$c = -\nabla_x \| y - f(x) \|_2^2 \tag{12}$$

The rest of the OMP and the CoSaMP algorithms remains as it is except for the least squares step. Previously, one needed to solve a linear least squares problem. We need to solve a nonlinear least squares problem. But solving nonlinear least squares is a well-studied area and there is no dearth of efficient algorithms.

Nonlinear OMP Algorithm:-

Initialize: Index set - $\Omega=\emptyset$,x=0
Repeat for k iterations
Compute correlation - c=$abs(\nabla_x \| y - f(x) \|_2^2)$
Select the index - i=$\underset{i}{\operatorname{argmax}} c_i$
Update support - $\Omega=\Omega \cup i$
Estimate nonzero values - $x_\Omega=\min_x \| y - f_\Omega(x) \|$
Finally, impute other indices in x with zeroes.

Nonlinear CoSaMP Algorithm:-

Initialize: Index set -$\Omega=\emptyset$,x=0
Repeat until convergence
Compute correlation - c=$abs(\nabla_x \| y - f(x) \|_2^2)$
Select the top 2 K support from c - $\omega$
Update support - $\Omega=\Omega \cup \omega$
Perform least squares estimate- $b_\Omega=\min_x \| y - f_\Omega(x) \|$
Update x by pruning b and keeping top K values.
Finally, impute other indices in x with zeroes.

We believe in reproducible research. MATLAB implementations of both the algorithms are available at the authors MATLAB Central account [9] free for use.

## 5 Experimental Evaluation

We test our nonlinear sparse recovery algorithms on two functions exponential (Ax) and logarithmic (A(c + x)). The Figs. 1 and 2 show the convergence of our modified OMP and CoSaMP algorithms for exponential and logarithmic functions, respectively, with 60 % sampling ratio and number of nonzeroes as 5. We want to study the variation of success rate with change in the number of nonzero values for a fixed number of measurements. An experiment is defined to be a success when the NMSE is below $10^{-3}$. Such definition of success rate is popular in CS literature [2, 3]. We keep the number of measurements (number of rows) fixed. The length of the signal x is 100, but the number of nonzero entries are now varied from 5 to 20 in steps of 5. The variation of success rate with varying number of nonzero values are shown in Figs. 3, 4, 5, and 6; Figs. 3 and 4 show the variation for exponential and logarithmic functions with 40 measurements and Figs. 5 and 6 show variations for 60 measurements.

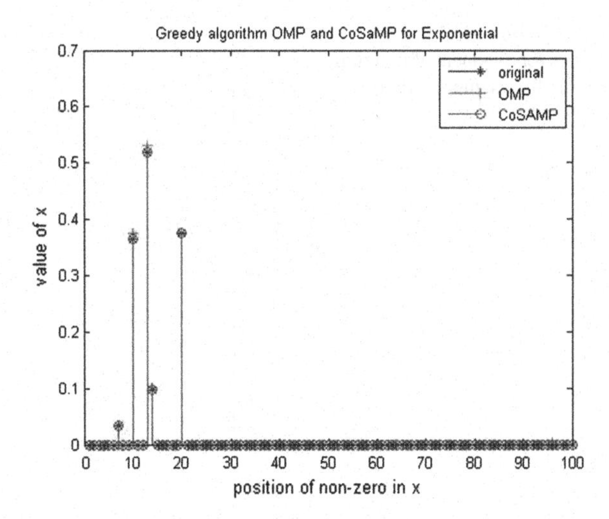


**Fig. 1** Convergence results for exponential function with 60 % sampling ratio and No. of nonzero as 5

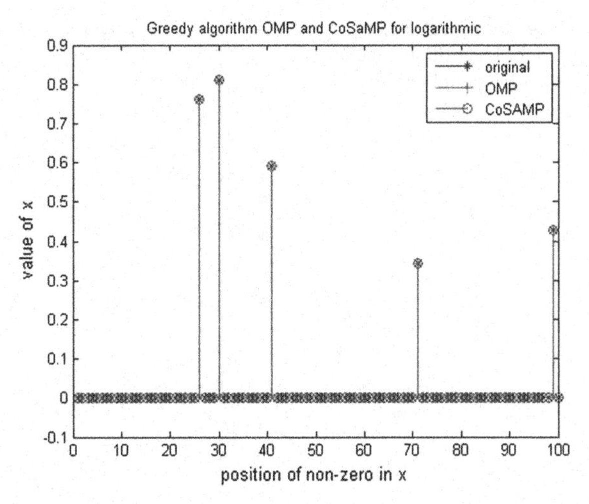


**Fig. 2** Convergence results for logarithmic function with 60 % sampling ratio and No. of nonzero as 5

To compute the empirical success rate each of the configurations are repeated 100 times and the average Normalized Mean Squared Error ($NMSE = \frac{\|original - reconstructed\|_2}{\|original\|_2}$) is reported.

There is hardly any work for nonlinear sparse recovery. Therefore, we do not have algorithms to benchmark against. In this work, we have used some standard evaluation criterions employed in CS to empirically study our algorithm. We find that

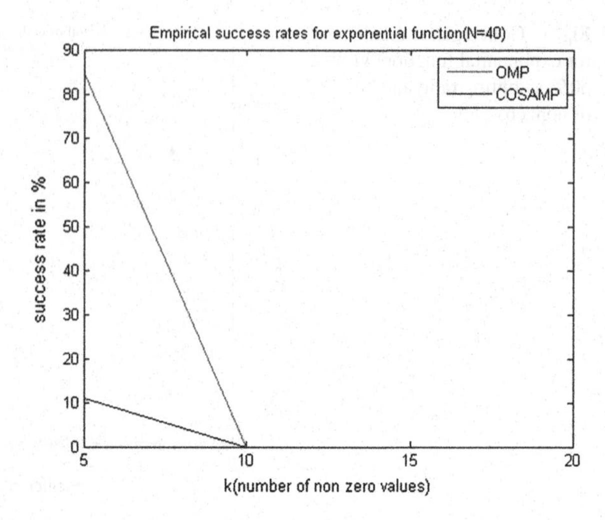


**Fig. 3** Empirical success rates for exponential function with 40 % sampling ratio

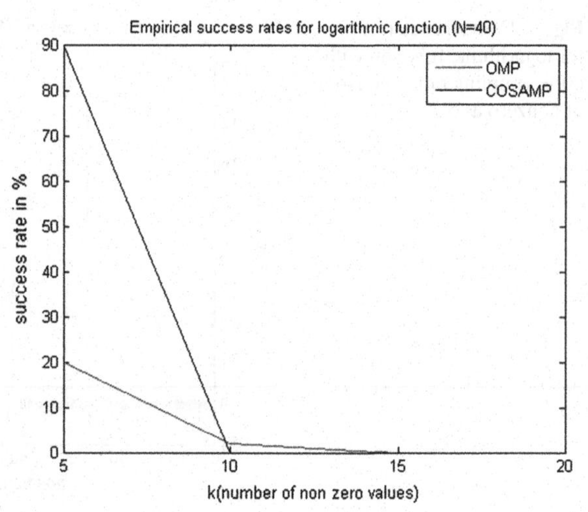


**Fig. 4** Empirical success rates for logarithmic function with 40 % sampling ratio

the OMP algorithm performs very well for the exponential function. The CoSaMP performs very well for the logarithmic function for higher sampling ratios; but the success rate drops down to zero when the number of nonzero values increase.

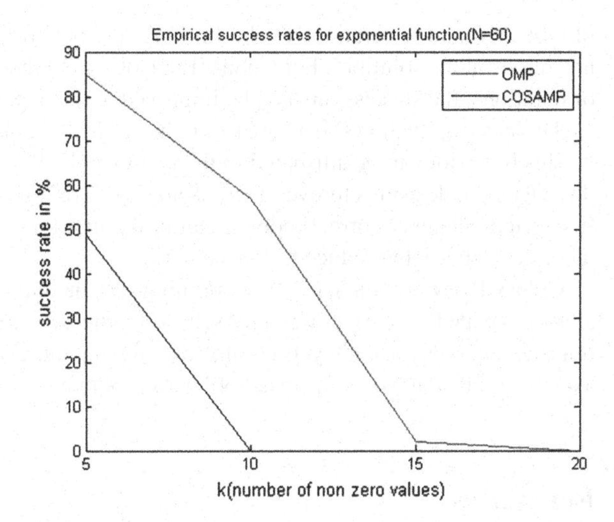


**Fig. 5** Empirical success rates for exponential function with 60 % sampling ratio

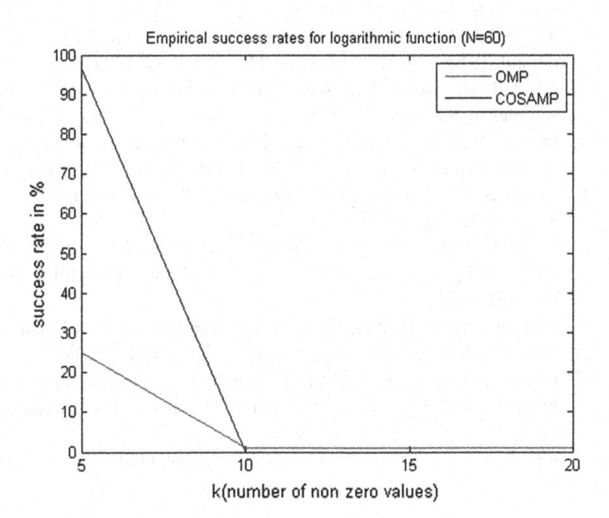


**Fig. 6** Empirical success rates for logarithmic function with 60 % sampling ratio

# 6 Conclusion

In this work, we modify two popular greedy sparse recovery algorithms to solve nonlinear problems. We use standard tests to evaluate the algorithms. We observed that the success rates are functions of the exit criteria for both the algorithms, especially the CoSaMP. In this work, we have used 2k iterations as the limiting number of CoSaMP and k iterations as the limiting number for OMP (k is the number of nonzeroes in the solution). For CoSaMP, we observed that by increasing the number of iterations, the success rate can be improved at the expense of larger run-times.

The success rate for OMP decreases if the number of iterations is increased beyond k. This behavior can be attributed to the fact that OMP does not allow pruning, i.e., the selected index stays forever; there is no way to remove a wrongly selected index. It selected whatever correct index it can in the first k iterations, after that it selects incorrect indices that reduce the success rate.

CoSaMP on the other hand allows pruning. The increase in the number of iterations allows CoSaMP to correct wrongly selected indices in previous passes; this in turn improves recovery accuracy. Basically, by allowing more and more iterations, we are asking CoSaMP to solve the exact NP hard problem.

## References

1. Donoho, D.L.: For most large underdetermined systems of linear equations the minimal $l_1$-norm solution is also the sparsest solution. Commun. Pure Appl. Math. **59**, 797–82 (2004)
2. Tropp, J.A., Gilbert, A.C.: Signal recovery from random measurements via orthogonal matching pursuit. IEEE Trans. Inf. Theory **53**, 4655–4666 (2007)
3. Needell, D., Tropp, J.A.: CoSaMP: iterative signal recovery from incomplete and inaccurate samples. arXiv:0803.2392
4. Das, P., Jain, M., Majumdar, A.: Non linear sparse recovery algorithm. In: IEEE International Symposium on Signal Processing and Information Technology, ISSPIT (2014)
5. Blumensath, T.: Compressed sensing with nonlinear observations. IEEE Trans. Inf. Theory **59**(6), 3466–3474 (2013)
6. Beck, A., Eldar, Y.C., Shechtman, Y.: Nonlinear compressed sensing with application to phase retrieval. In: Global Conference on Signal and Information Processing (GlobalSIP, 2013), p. 617. IEEE (2013)
7. Blumensath, T., Davies, M.E.: Gradient pursuit for non-linear sparse signal modelling. In: European Signal Processing Conference (EUSIPCO) (2008)
8. Donoho, D.L., Tsaig, Y.,Drori, I., Starck, J.-L.: Sparse solution of underdetermined linear equations by stagewise orthogonal matching pursuit. IEEE Trans. Inf. Theory **58**(2), 1094–1121 (2012)
9. http://www.mathworks.in/matlabcentral/fileexchange/48285-greedy-algorithms-for-non-linear-sparse-recovery

# Reducing Inter-Scanner Variability in Multi-site fMRI Activations Using Correction Functions: A Preliminary Study

**Anoop Jacob Thomas and Deepti Bathula**

**Abstract** In the past decade, there has been an exponential growth in brain mapping studies using Functional MRI (fMRI). The need for more data to increase the statistical power of brain mapping studies made the researchers look at multi-center studies. But a major limitation in pooling data from multiple sites is the diversity in acquisition and analysis methods that effect the imaging results. This preliminary study aims at finding correcting functions to reduce the variability of activation maps produced using fMRI for a particular task at different sites. Having explored corrections based on ordinary least squares (OLS) and robust estimation methods, we found that even simple linear correction functions produced reasonable reduction in variability of activation maps across sites.

**Keywords** Multi-center · Multi-site · fMRI · Inter-scanner · Variability · Correction functions

## 1 Introduction

Functional Magnetic Resonance Imaging (fMRI) is a relatively safe medical imaging technique. fMRI measures the blood oxygenation levels during a scan period producing BOLD (Blood Oxygenation Level Dependent) time series data for the individual voxels of the brain. In clinical application, fMRI is used for brain function mapping, used for detecting neuro-degenerative diseases and for identifying neurological disorders. Data from different demographics is required such that the study is not biased by a population sample, thus data pooling becomes an important aspect in fMRI studies. Since fMRI is a relatively new technique and due to the cost associated

A.J. Thomas (✉) · D. Bathula
Department of Computer Science and Engineering, Indian Institute of Technology Ropar, Rupnagar, Punjab 140001, India
e-mail: anoopjt@iitrpr.ac.in

D. Bathula
e-mail: bathula@iitrpr.ac.in



R. Singh et al. (eds.), *Machine Intelligence and Signal Processing*, Advances in Intelligent Systems and Computing 390,
DOI 10.1007/978-81-322-2625-3_10

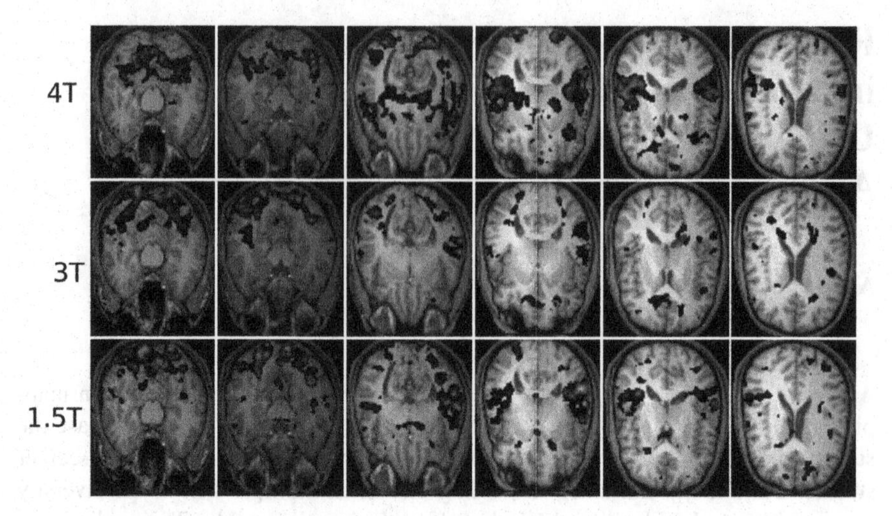


**Fig. 1** Activation map showing the functionally activated regions for sensorimotor task of a single subject at three different scanners (rows). The activation regions are overlaid on high-resolution anatomical images for better visualization

with the scans, diverse data from different population samples is not available and thus it poses a problem in functional brain mapping. Brain is a complex organ, and thus it is not possible to conclude on the brain function based on a single scan from a single subject. Data from multiple subjects and multiple scanners can be combined to increase the statistical confidence of the brain mapping studies [1, 7]. But, adding to the complexity of the organ under study, the scanner hardware heterogeneity and scanner software heterogeneity also affect the data. fMRI data thus acquired at different scanners for the same subject performing the same activity is not exactly the same, and thus the activation maps also look different (Fig. 1). Reducing inter-scanner variability in multi-center fMRI data thus becomes an important consideration for the purpose of data pooling. In this paper, we explore the use of correction functions to reduce the inter-scanner variability of fMRI data.

## 2 Related Work

Data pooling is possible only if the between-site reliability of the multi-site fMRI data is high. Classically, reliability is assessed with ICC [5] (intra-class correlation coefficient) which is estimated using the variance components. One way to pool the multi-site fMRI data would be to consider the data originated from a single site and perform the analysis, but accept the result if and only if the reliability of the fMRI data between sites is high (site is considered a variable). In a similar study, Friedman et al. [5] point out the low between-site reliability, due to a strong contribution from site and site-by-subject variance.

Using methods to increase the between-site reliability of fMRI data is another way to achieve pooling. Friedman et al. in two of their studies try to reduce the inter-scanner variability in multi-center fMRI data, one investigating the role of smoothness equalization [3], and another by controlling for the SFNR (signal-to-fluctuation-noise ratio) [4]. Smoothness equalization requires the dataset to be smoothed to the smoothest data. In the paper on controlling SFNR, the PVAF (Percent of Variance Accounted For) maps showed marked scanner effects and variance components analyses revealed significant reductions in variance due to scanner.

Structural MR imaging is also affected by the scanner-to-scanner variation. Kelemen et al. [8] explored the use of correction functions to reduce the differences of structural MR images from multi-center MRI studies. It was shown that the between-scanner differences were higher than the between-subject variance. Inspired by the idea of using correction functions, we explore the use of correction functions for multi-center fMRI data.

# 3 Materials and Methods

## *3.1 Data*

This study uses fBIRN Phase I Traveling Subjects dataset. Five healthy, right-handed, and English-speaking male subjects were imaged at least on two visits on ten (nine sites) different scanners (vendors, field strength) located in geographically diverse locations within the United States. Each subject visited every site at least twice, with few subjects visiting few of the sites four times. During each of the visits on consecutive days, the subjects were scanned for multiple tasks, and the tasks included resting state fMRI scan, breath-hold task, working memory task, attention task, and four runs of sensorimotor task per visit. Along with BOLD-based functional image acquisitions, high-resolution 3D anatomical images (T2-weighted, fast spin-echo, turbo factor = 12 or 13, orientation: parallel to the AC-PC line, number of slices = 35, slice thickness = 4 mm, no gap, TR = 4000 ms, TE = approximately 68, FOV = 22 cm, matrix = 256 $\times$ 192, voxel dimensions = 0.86 mm $\times$ 0.86 mm $\times$ 4 mm) were also acquired during every visit for mapping the functional activity.

The functional data were collected using echo-planar (EPI) trajectories (7 scanners) or spiral trajectories (3 scanners) (orientation: parallel to the AC-PC line, number of slices = 35, thickness = 4 mm, no gap, TR = 3000 ms, TE = 30 ms on the 3 and 4 T scanners, 40 ms on the 1.5 T scanners, FOV = 22 cm, matrix = 64 $\times$ 64, voxel dimensions = 3.4375 mm $\times$ 3.4375 mm $\times$ 4 mm). One of the 3 T scanners employs a double echo EPI sequence. The scanners were from GE, Siemens, and Picker, with one 4-T scanner, four 3-T scanners, and five 1.5-T scanners, with similar scanning protocols.

## 3.2 Sensorimotor Task

The sensorimotor task was designed as a block design, with 15 s of rest block followed by 15 s of sensorimotor activity block, with a total of eight such cycles followed by 15 s of rest. The task measured a total of 85 time points with a TR of 3 s (4.25 min). Task included stimulation of the auditory, visual, and sensorimotor regions of the brain. During the sensorimotor activity block, the subjects were shown a 3 Hz flashing checkerboard, a 3 Hz midi-tone simultaneously, and the subjects were asked to perform alternating finger tapping (index, middle, ring, little, little, ring, middle, index, index…) (not using thumb) according to the simultaneous audio and visual stimulus. During the rest block, the subjects were asked to stare at a fixation cross. Due to pre-processing related issues and missing data, our study uses a total of eight scanners and four subjects, with one 4-T scanner, three 3-T scanners, and four 1.5-T scanners.

## 3.3 Methods

Individual sensorimotor runs for each of the subjects for each of the visits to different scanners were first subjected to six different pre-processing steps, and then three different levels of subject-wise analysis (refer Fig. 2). The pre-processing steps are

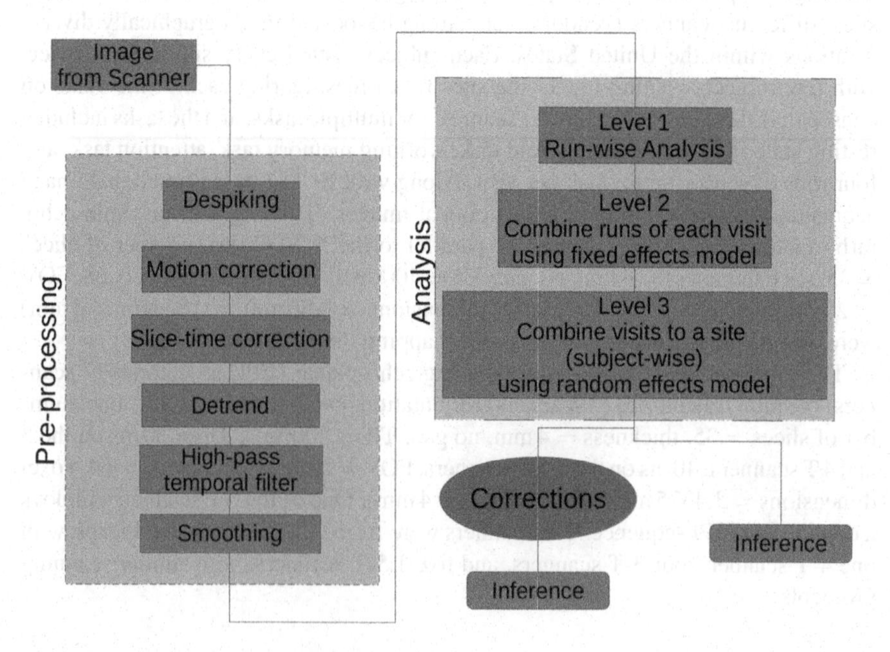


**Fig. 2** Schematic diagram of the fMRI processing

required to correct for acquisition related artifacts, physiological noise, system related noise, and assumptions about data.

Each sensorimotor run undergoes a series of pre-processing steps in the order, despiking, motion correction, slice-time correction, detrending with a second-order polynomial, temporal high-pass filtering and smoothing with a $7 \times 7 \times 7$ kernel. The pre-processing steps were decided based on the studies by Friedman et al. [3, 4], which used the same dataset.

The pre-processing pipeline was implemented using Nipype [6], using different modules from different fMRI processing softwares. Despiking is accomplished using AFNI [2]. Motion correction is done using SPM8 [10], motion correcting the functional scans to a mean functional image constructed for each subject, so as to spatially align the scans of each subject. Slice-time correction is done using SPM8. Detrending is carried out with a second-order polynomial using AFNI. Temporal high-pass filtering is accomplished using FSL [9], using a lower cut-off frequency of 0.02 Hz to remove the physiological noise like cardiac and respiration. Finally, the data is smoothed using a $7 \times 7 \times 7$ mm kernel using SPM8 to increase the signal-to-noise ratio (SNR).

Three levels of analysis are performed as part of combining the data to increase the statistical confidence. After pre-processing, each individual run is subjected to a level 1 fMRI analysis using GLM (General Linear Model), in which the block design is convolved with a double gamma function to form the first predictor in the design matrix and a univariate analysis is done for each of the individual voxel time series. The level 1 data from each visit for each of the subjects to a scanner is combined using fixed-effects model in level 2 analysis, combining the four runs of a single visit, producing visit-wise fMRI activation maps per subject per visit to scanner. Each subject has at least two visit activation maps per scanner and utmost four visit activation maps. In level 3 analysis, the data from the visits of a subject to a scanner is combined using the random-effects model, forming one activation map per subject per scanner.

Level 1, level 2, and level 3 analyses were accomplished using FMRISTAT software by Keith J Worsley [11]. The analysis produced t-stat images which were used to estimate the correction functions.

## 4 Estimating the Correction Functions

For estimating the corrections, it is very important to select a reference image to which an image from another scanner can be corrected to. In this study, we use the highest field strength scanner, the 4-T scanner, as the reference image and try to correct the images from the 3-T scanners and the 1.5-T scanners to the 4-T scanner. Zou et al. [12] showed that higher field strength magnets have better reproducibility of activations across sites and significantly better results than 1.5-T imaging reproducibility; the lowest field strength magnet in the study was 1.5-T.

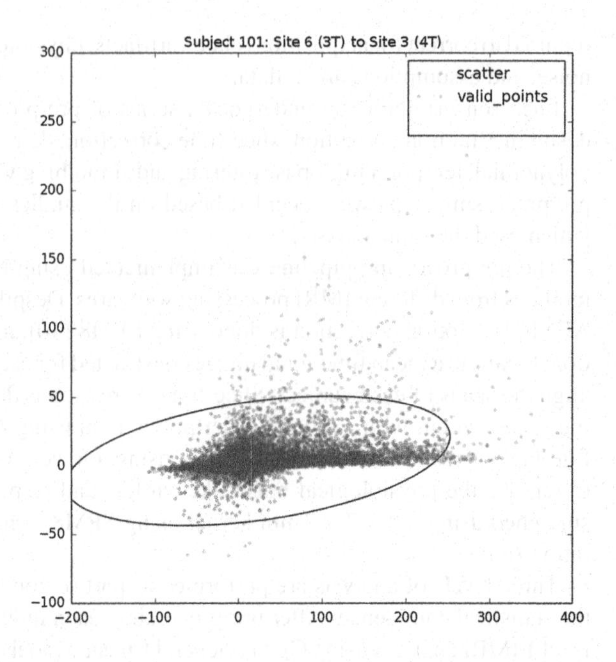


**Fig. 3** A sample scatter plot showing the density ellipse-based robust estimation with standard deviation of 4.5

The correction functions were estimated using the voxels from the brain region. The mask for the whole brain region is created by making binary image from skull-stripped mean functional image for each subject. The corrections were estimated using scatter plots from uncorrected image t-statistics value and reference image t-statistic values. The uncorrected and reference images were arranged in abscissa and ordinate, with the t-statistic values arranged into a vector according to spatial correspondence. The relationship was estimated using ordinary least squares (OLS) and robust estimation methods.

Two different robust estimations were tried: density ellipse-based outlier elimination (refer Fig. 3) and an in-house method (named EOEL, experimental outlier elimination), which marks points outside a certain aggregate standard deviation as outliers. The final threshold points for EOEL are found by fitting a degree two curve through the end points selected at a distance of a definite standard deviation from mean. The EOEL method removes less outliers as compared to the density ellipse method.

## 5 Results and Discussion

The correction with OLS increases the sensitivity at a false positive rate of 20–25 %, and the sensitivity increases from 40 % to 75–80 % (refer Fig. 4). But the between-subject differences seem to play an effect. We also tried robust estimation methods to estimate the corrections by removing the outliers. But we observe that

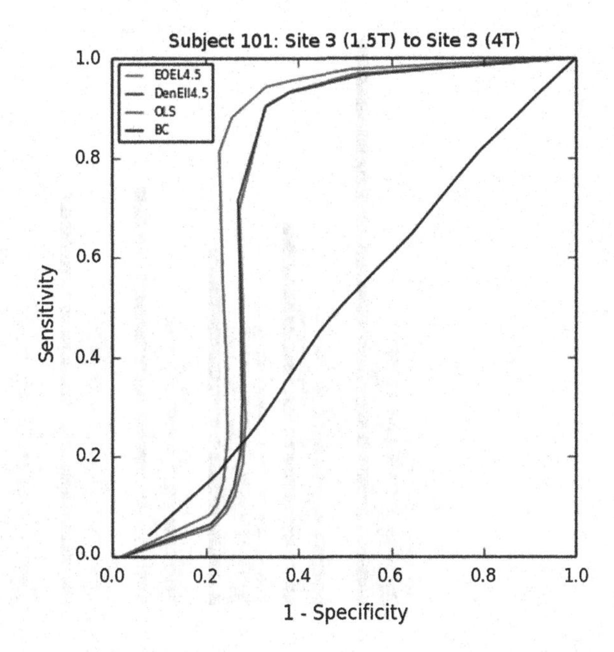


**Fig. 4** (1-Specificity) versus Sensitivity plot of Site 3 (1.5-T) scanner corrected to 4-T scanner, showing an average profile of before correction (BC in *black color*) and after corrections with various correction functions

the robust estimation methods do not provide any significant improvement over OLS estimation. The outliers from the robust estimations were observed toward the tissue boundaries and it could be attributed to brain pulsations due to cerebral blood flow and partial volume effects due to lower spatial resolution in fMRI. The outlier elimination methods mark 1–2 % of the points as outliers.

The mean square error shows marked improvement after correction (refer Fig. 5). Although the robust estimation methods also show a relatively good improvement in mean square error after the correction, the (1-Specificity) versus Sensitivity (FPR versus TPR) shows that the OLS performs better than the robust methods. Figure 6 shows the activation maps after correction, thresholded at $p < 0.05$, with 4 T map as the ground truth, marking true positives (in blue color), false positives (in red color), and false negatives (in green color).

We explored the possibility of higher order correction functions upto degree eight but no significant improvement over linear corrections has been observed. The goodness of fit of the linear estimates is low (~0.05–0.25), and thus we can further explore more complicated corrections using neural networks to uncover the underlying structure of the data. We can also opt for data reduction techniques to reduce the complexity of the data by reducing the number of points from which the corrections are to be estimated.

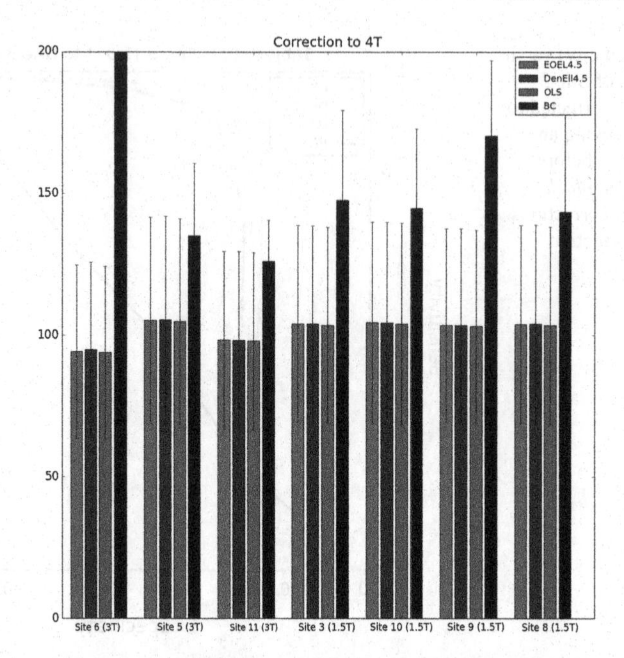


**Fig. 5** Mean square error before correction (BC in *black color*) and after correction, with each block representing a scanner corrected to 4 T reference scanner

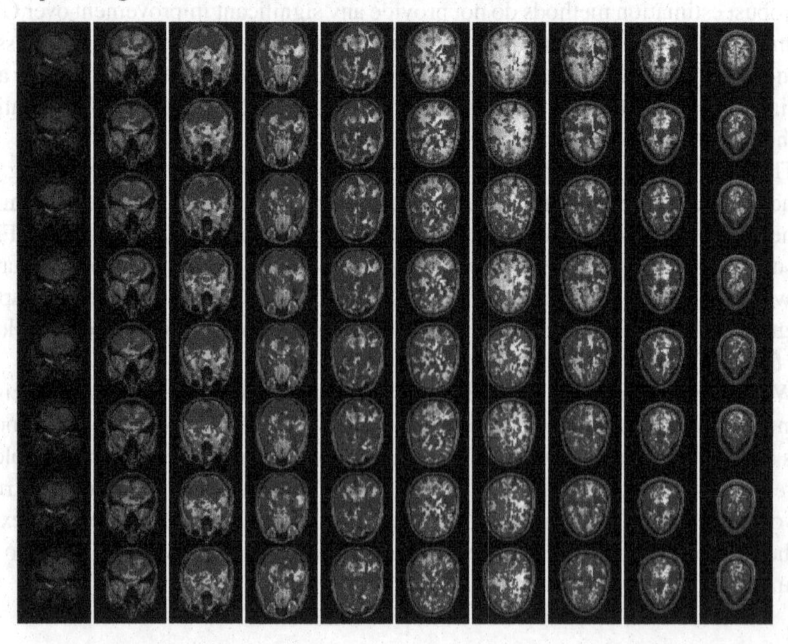

**Fig. 6** A sample activation map (sub 106) showing the true positives in the *blue color*, the false positives in *red*, and the false negatives in *green* with respect to the 4 T activation map. The rows are Site 3 (4 T), Site 6 (3 T), Site 5 (3 T), Site 11 (3 T), Site 3 (1.5 T), Site 10 (1.5 T), Site 9 (1.5 T), and Site 8 (1.5 T)

**Acknowledgments** Some of the data used in this study was acquired through and provided by the Biomedical Informatics Research Network under the following support: U24-RR021992, Function BIRN and U24 GM104203, Bio-Informatics Research Network Coordinating Center (BIRN-CC).

## References

1. Costafreda, S.G.: Pooling fMRI data: meta-analysis, mega-analysis and multi-center studies. Front. Neuroinf. **3**, 33 (2009). http://www.frontiersin.org/neuroinformatics/10.3389/neuro.11/033.2009/abstract
2. Cox, R.W.: Afni: software for analysis and visualization of functional magnetic resonance neuroimages. Comput. Biomed. Res. **29**(3), 162–173 (1996)
3. Friedman, L., Glover, G.H., Krenz, D., Magnotta, V.: Reducing inter-scanner variability of activation in a multicenter fMRI study: role of smoothness equalization. NeuroImage **32**(4), 1656–1668 (Oct 2006). http://www.sciencedirect.com/science/article/pii/S1053811906004435
4. Friedman, L., Glover, G.H., The FBIRN Consortium: Reducing interscanner variability of activation in a multicenter fMRI study: controlling for signal-to-fluctuation-noise-ratio (SFNR) differences. NeuroImage **33**(2), 471–481 (Nov 2006). http://www.sciencedirect.com/science/article/pii/S1053811906007944
5. Friedman, L., Stern, H., Brown, G.G., Mathalon, D.H., Turner, J., Glover, G.H., Gollub, R.L., Lauriello, J., Lim, K.O., Cannon, T., Greve, D.N., Bockholt, H.J., Belger, A., Mueller, B., Doty, M.J., He, J., Wells, W., Smyth, P., Pieper, S., Kim, S., Kubicki, M., Vangel, M., Potkin, S.G.: Test-retest and between-site reliability in a multicenter fMRI study. Hum. Brain Mapp. **29**(8), 958–972 (Aug 2008). http://www.ncbi.nlm.nih.gov/pmc/articles/PMC3670112/
6. Gorgolewski, K., Burns, C.D., Madison, C., Clark, D., Halchenko, Y.O., Waskom, M.L., Ghosh, S.S.: Nipype: a flexible, lightweight and extensible neuroimaging data processing framework in python. Front. Neuroinf. **5** (2011)
7. Gountouna, V.E., Job, D.E., McIntosh, A.M., Moorhead, T.W.J., Lymer, G.K.L., Whalley, H.C., Hall, J., Waiter, G.D., Brennan, D., McGonigle, D.J., Ahearn, T.S., Cavanagh, J., Condon, B., Hadley, D.M., Marshall, I., Murray, A.D., Steele, J.D., Wardlaw, J.M., Lawrie, S.M.: Functional magnetic resonance imaging (fMRI) reproducibility and variance components across visits and scanning sites with a finger tapping task. NeuroImage **49**(1), 552–560 (Jan 2010). http://www.sciencedirect.com/science/article/pii/S1053811909007988
8. Kelemen, A., Liang, Y.: Multi-center correction functions for magnetization transfer ratios of mri scans. J. Health Med. Inf. **0**(0)
9. Smith, S.M., Jenkinson, M., Woolrich, M.W., Beckmann, C.F., Behrens, T.E., Johansen-Berg, H., Bannister, P.R., De Luca, M., Drobnjak, I., Flitney, D.E., et al.: Advances in functional and structural MR image analysis and implementation as FSL. Neuroimage **23**, S208–S219 (2004)
10. Worsley, K.J., Friston, K.J.: Analysis of FMRI time-series revisited again. Neuroimage **2**(3), 173–181 (1995)
11. Worsley, K., Liao, C., Aston, J., Petre, V., Duncan, G., Morales, F., Evans, A.: A general statistical analysis for fMRI data **15**(1), 1–15 (2002). http://www.sciencedirect.com/science/article/pii/S1053811901909334
12. Zou, K.H., Greve, D.N., Wang, M., Pieper, S.D., Warfield, S.K., White, N.S., Manandhar, S., Brown, G.G., Vangel, M.G., Kikinis, R., Wells, W.M.: Reproducibility of functional MR imaging: Preliminary results of prospective multi-institutional study performed by biomedical informatics research network1. Radiology **237**(3), 781–789 (Dec 2005). http://radiology.rsna.org/content/237/3/781

# Comparative Study of Preprocessing and Classification Methods in Character Recognition of Natural Scene Images

**Yash Sinha, Prateek Jain and Nirant Kasliwal**

**Abstract** This paper presents an approach to character recognition in natural scene images. Recognizing such text is a challenging problem in the field of Computer Vision, more than the recognition of scanned documents due to several reasons. We propose a classification technique for classifying characters based on a pipeline of image processing operations and ensemble machine learning techniques. This pipeline tackles problems where Optical Character Recognition (OCR) fails. We present a framework that comprises a sequence of operations such as resizing, grey scaling, thresholding, morphological opening and median filtering on the images to handle background clutter, noise, multi-sized and multi-oriented characters and variance in illumination. We used image pixels and HOG (Histogram of Oriented Gradients) as features to train three different models based on Nearest-Neighbour, Random Forest and Extra Tree classifiers. When the input images were pre-processed, HOG features were extracted and fed into extra tree classifier, and the model classified the characters with maximum accuracy, among the other models that we tested. The proposed steps have been experimentally proven to yield better accuracy than the present state-of-the-art classification techniques on the Chars74k dataset. In addition, the paper includes a comparative study elaborating on various image processing operations, feature extraction methods and classification techniques.

**Keywords** Camera-based character recognition · Histogram of oriented gradients · Feature extraction · Scene text recognition · Ensemble classifiers

## 1 Introduction

We focus on recognition of individual characters from natural scene images. It is a field of active research with high utility and potentially on high-impact applications. Unlike scanned greyscale images, natural scene images have characters with additional colour information and noise.

Y. Sinha (✉) · P. Jain · N. Kasliwal
Department of Computer Science and Information Systems,
Birla Institute of Technology and Science, Pilani 333031, India
e-mail: mail.yash.sinha@gmail.com



R. Singh et al. (eds.), *Machine Intelligence and Signal Processing*,
Advances in Intelligent Systems and Computing 390,
DOI 10.1007/978-81-322-2625-3_11

The challenge arises due to variation in fonts, styling, formatting, background clutter, size and orientation of characters in addition to background noise and occlusion. Additionally, low resolution, uneven illumination, glare, gloss and artistic format make the classification error prone. These are the reasons that commercial OCR engines have dismal performance on natural scene images.

Our proposed methodology tackles these challenges by binarization, morphological opening and feature extraction. The extracted features are used as input for the classification models. The 62 classes into which the images can be categorized are 0–9, a–z and A–Z.

The performance and accuracy achieved are far better than optical character recognition systems. This yielded better accuracy than the present state-of-the-art technique. [1]

## 2 Related Work

Recent papers on character recognition in natural scenes have proposed multiple techniques of image processing, feature extraction and classification models. The problem lies on the common intersection of computer vision and machine learning. Campos et al. [2] introduced Chars74k dataset of characters collected from natural scene images. The work highlighted that commercial OCR engines do not provide satisfactory results for natural scene character images. Their experiment was based on methods of Multiple Kernel Learning, Nearest-Neighbour Classification and Support Vector Machines. Sheshadri et al. [3] proposed the exemplar-based multi-layered classification approach to recognize characters. They proposed a methodology using SVM to classify images on features from Histogram of Oriented Gradients (HOG) [4] on resized and warped images. Fraz et al. [1] have recently proposed a pipeline of image processing operations for the identification of characters. Zhang et al. [5] used a mixed dataset of natural scene images, handwritten characters and digital text images to benchmark performance for parametric nearest-neighbour technique. They explored efficacy of colour information and variation for the identification of characters from an image to give accuracy of 72 % for 49 classes. Kumar et al. [6] applied Discrete Cosine Transform (DCT) for character recognition after binarization to remove noise for English and Kannada character recognition. kNN was used as the classification model. Neumann et al. [7] achieved a word recognition rate of 72 % using graph-based approaches in Maximally Stable Extremal Regions (MSER). Machine learning model used was a SVM with RBF. Proposed methods focus on individual character recognition with techniques for image processing, feature extraction and machine learning techniques on the images.

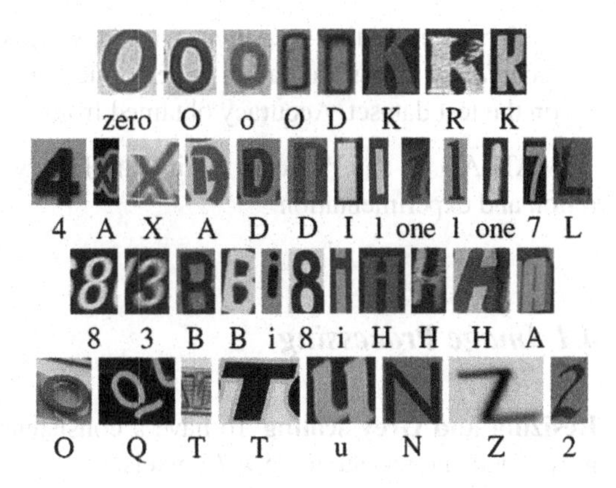


**Fig. 1** Examples of images from dataset

## 3 Dataset

The data is taken from the Chars74K dataset [8], which consists of images of characters selected from Google Street View images *(EngImg)*, as shown in Fig. 1. Street view images consist of English characters taken in natural contexts such as street signs, house names and numbers, shop signboards and so on. The dataset differentiates between digits, uppercase and lowercase characters to give a total of 62 classes. Individual characters were manually segmented from these images. Our dataset has 12,503 characters, of which 6283 (50.25 %) are randomly sampled for training, while rest 6220 (49.75 %) are used as test data. There is significant variation in thickness, size, alignment, orientation and position of characters both within the class and across the classes. Handwritten and synthetic images were not used.

## 4 Experiments and Results

**Proposed Methodology**: The proposed methodology consists of four steps:

1. Image processing—*tackles background clutter, fonts and illumination of images*: This step includes converting all images to a common standard size, binarization using Otsu's Thresholding followed by Morphological opening and median filter for removing salt and pepper noise.
2. Feature extraction—*addresses variation in character size and orientation*: Histogram of Oriented Gradients is used as a method of feature extraction. The value of used cell size is calibrated experimentally to an optimal value. It also reduces the number of features in comparison to using direct pixel values.
3. Training the classifier—*does the learning for classifier models and classifies test images into classes*: kNN, random forest and extra tree classifier are tested and compared. The number of estimators is determined experimentally to an optimal value. The input parameters HOG features and pixel values both are tested and compared.

4. Performance testing—*compares various classifications and feature extraction techniques*: Cross validation is performed and the classification model is tested on the test dataset. Accuracy obtained from both the methods is mentioned.

MATLAB [9] and Python's scikit-learn library [10] are used for the above exploration and experimentation.

## *4.1 Image Processing*

**Resizing and Grey scaling** To have a consistent feature set, all the images were greyscaled and resized to 75 × 75 pixels.

The work of Fraz demonstrated that processing colour information is an additional challenge. In order to tackle this, images were greyscaled using a weighted sum of the R, G and B components.

The images were enlarged using bi-cubic interpolation, where the output pixel value is a weighted average of pixels in the nearest 4-by-4 neighbourhood. Enlarging the image led to extraneous edge detection. Accuracy was improved because more features increased variability in the data.

The size of images was reduced using antialiasing. Visual inspection showed that reducing the image size led to noise suppression in ensuing edge detection.

**Thresholding** Otsu's method for thresholding [11] is an image processing technique for image segmentation. The algorithm calculates the optimum threshold separating the two classes (foreground/object and background) so that their combined spread is minimal. Separating the foreground and background pixels is necessary due to high variance and noise in the data set.

We tried basic thresholding, average thresholding and Otsu thresholding techniques (Fig. 2). In comparison with average thresholding, we observed that Otsu retains much more information about local changes in the pixel values.

**Morphological Operations** Morphological operations rely only on the relative ordering of pixel values and are especially suited to the processing of binary images.

The image was opened [12] using a disc of 1 pixel radius. In our particular case, this removed the salt and pepper noise in addition to smoothing the edge and filling the gaps within the character strokes (leaving the holes unfilled) (Fig. 3).

**Fig. 2** Comparison of thresholding techniques: No thresholding, Average thresholding, Otsu thresholding

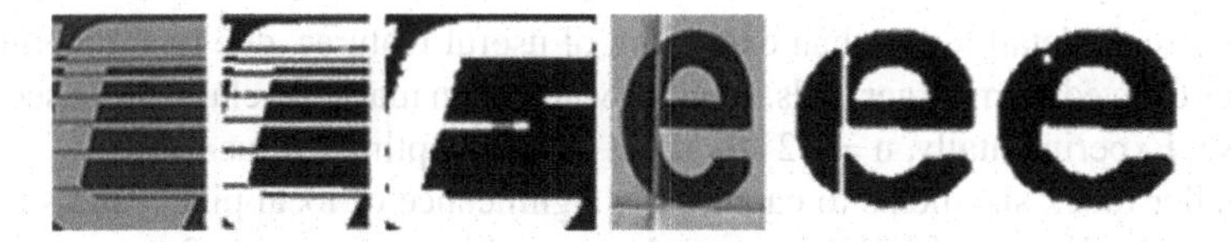

**Fig. 3** Comparison of morphological operations: Original image, Average thresholding, Morphological Opening

**Noise Reduction** Median filter showed better results than the averaging filter, as expected [13] for noise reduction. Median filter yields enhanced performance because it preserves edge details better than mean filter. It allows high spatial frequency detail to pass while remaining very effective at removing noise [14].

## *4.2 Feature Extraction*

**Image Pixels** We represented the 2D image matrix as single-dimensional vector with pixels as features for classification. This gave 5625 features.

**Histogram of Oriented Gradients** HOG creates histograms of edge orientation to compute the gradients of the edges from patches of the image. HOG as feature extraction technique is independent of variation in orientation, and hence is advantageous.

HOG is an iterative process, where a cell of n × n pixels is extracted from the image and algorithm is applied to calculate the histogram of oriented gradients. These gradient values are used as features of the image.

A series of values for n were applied. In HOG, with increasing value of n, there is a decrease in the number of features in the image (Fig. 4). To capture large-scale spatial information, we increased the cell size. However, for n > 12, there was loss

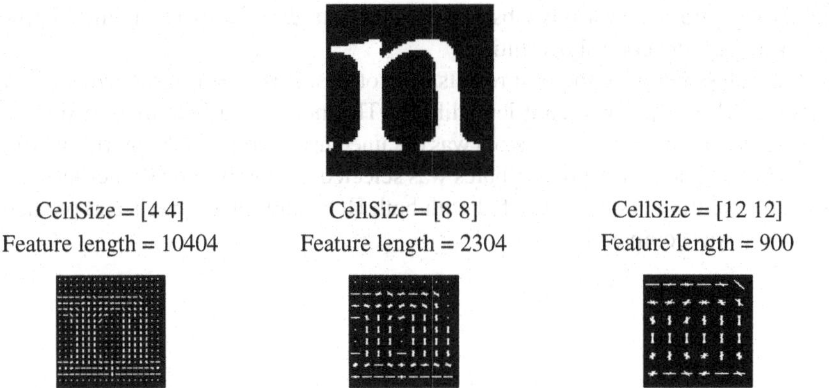


**Fig. 4** Visualization of HOG features

of small-scale detail rather than extraction of useful features, due to submerging of features derived from larger cells. Smaller values of n make the classifier susceptible to noise. Experimentally, $n = 12$ stood out to be an optimal value to use.

Smaller block size helps to capture the significance of local pixels and suppress illumination changes of HOG features. Hence, a block of size $2 \times 2$ was used.

To ensure adequate contrast normalization, an overlap of at least half the block size: $1 \times 1$ was selected. Large overlap values can capture more information, but they produce larger feature vector size.

To achieve a trade-off between finer orientation details and size of the feature vector, the number of orientation histogram bins was selected as 9. This gave 900 features.

### 4.3 Classification Techniques

**K-Nearest Neighbour (kNN)** Taking the one-dimensional n bit vector image as a point in n-dimensional hyperspace, we applied kNN [15] for classification using *neighbours* sub-module of Python's scikit-learn library [16]. Value of k was varied from 1 to 9 to select appropriate k value.

The kNN model used for both HOG and Image pixels used Euclidean distance to calculate distance. The weight for all k obtained points in kNN had uniform weights.

**Random Forest Classifier** It is a meta-estimator [17] based on subsampling which controls overfitting well. The *ensemble* sub-module of Python's scikit-learn library was used [18].

The model computed 1025 trees (estimators), using GINI criterion for splitting tree nodes and applied ensemble technique to make final prediction. The value for number of estimators was obtained using grid search [19]. It was fed with both HOG and Image pixels and prediction accuracy was computed.

**Extra Tree Classifier** An estimator [20] fits a number of randomized decision trees (a.k.a. extra trees) on various subsamples of the dataset and uses averaging to improve the accuracy and control overfitting.

This classifier gave the best results over others. It is a part of the *ensemble* sub-module [21] of Python's scikit-learn library. The model used 1025 trees (estimators). The value for number of estimators was obtained using grid search algorithm [19]. In extra tree classifier, a set of attributes was selected randomly and GINI criterion was used for splitting nodes. It was fed with both HOG and Image pixels and prediction accuracy was computed.

### 4.4 Performance Testing

The Chars74k dataset was partitioned into two sets. The model was trained over 50.25 % of the dataset. The rest of the images were used to test the model and calculate the estimator performance. We performed 10-fold cross validation on training set to improve on training efficiency.

### 4.5 Results

We fed the features obtained from HOG (preceded by other processing) into extra tree classifier and tested our model using tenfold cross validation and test data. Our proposed pipeline accurately classified 73.167 % (Table 2) of the character images in the test dataset, which beats the current state-of-the-art technique [5] as shown in Table 1. On cross validation, 73.022 % accuracy was obtained as shown in Table 3.

**Table 1** Accuracy comparison with other methods

| Method | Accuracy (%) | Dataset | Classes |
|---|---|---|---|
| Proposed method | 73.167 | Chars74K | 62 |
| Fraz et al. [1] | 72 | Chars74K-15 | 49 |
| Shi et al. [22] | 69.9 | Chars74K-15 | 62 |
| Newell et al. [23] | 66.5 | Chars74K-15 | 62 |
| Campos et al. [2] | 55.26 | Chars74k | 62 |

**Table 2** Accuracy across different features and classifiers on test data

| Features | kNN(k = 4) (%) | Random forest (%) | Extra tree (%) |
|---|---|---|---|
| HOG features | 69.549 | 71.575 | **73.167** |
| Image pixels | 55.659 | 62.636 | 64.244 |

**Table 3** Cross validation scores across different features and classifiers

| Features | kNN(k = 4) (%) | Random forest (%) | Extra tree (%) |
|---|---|---|---|
| HOG features | 69.220 | 71.590 | **73.022** |
| Image pixels | 55.120 | 62.013 | 63.570 |

## 5 Observations and Trends

**A Comparative Study**:

A comparative study elaborating on various image processing operations, feature extraction methods and classification techniques is presented here. Certain trends of the exploration have been discussed.

### *5.1 Image Pixels as Feature Set Versus HOG Features*

Edge orientations of images give better features to the training model as compared to image pixels. HOG features help reduce variations arising from orientation and size despite sensitivity to non-random noise. While using HOG features with extra tree classifier, the accuracy increased by 10 % on 10-fold cross validation scores and 9 % for test data (Fig. 5).

### *5.2 Average Thresholding Versus Otsu Thresholding*

Different thresholding algorithms approach noise removal in images differently. Otsu's thresholding improved the accuracy in comparison to average thresholding by retaining more information in high noise environments (Fig. 6).

### *5.3 K-Nearest Neighbours (1NN–9NN)*

The graph shows scores of K-Nearest Neighbours classifier with different values of k on test data. Clearly, 1NN and 4NN outperform other k values (Figs. 7 and 8).

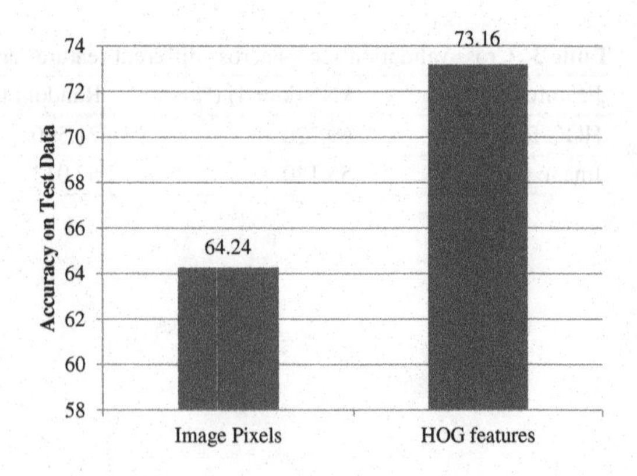


**Fig. 5** Image Pixels as feature set Versus HOG features

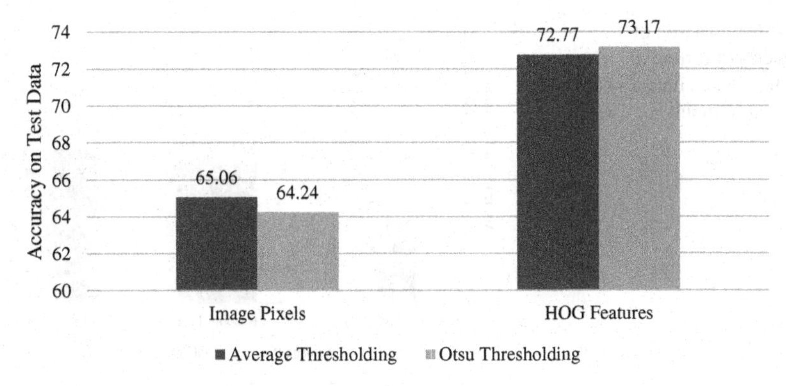


**Fig. 6** Thresholding trends on image pixels and HOG features as feature set

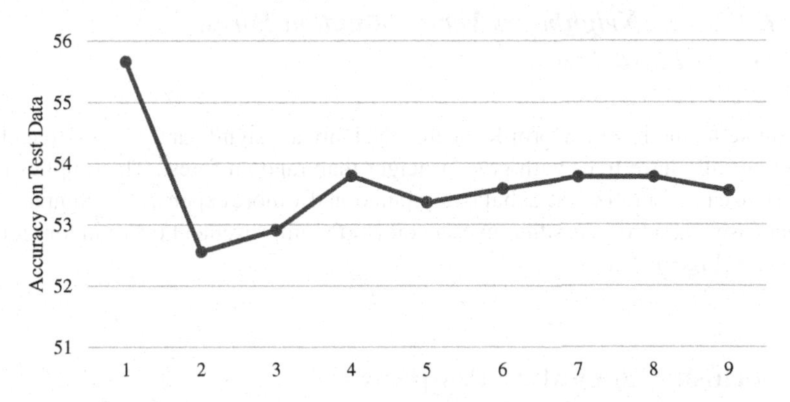


**Fig. 7** K-Nearest Neighbours with different values of k on Image Pixels as feature set

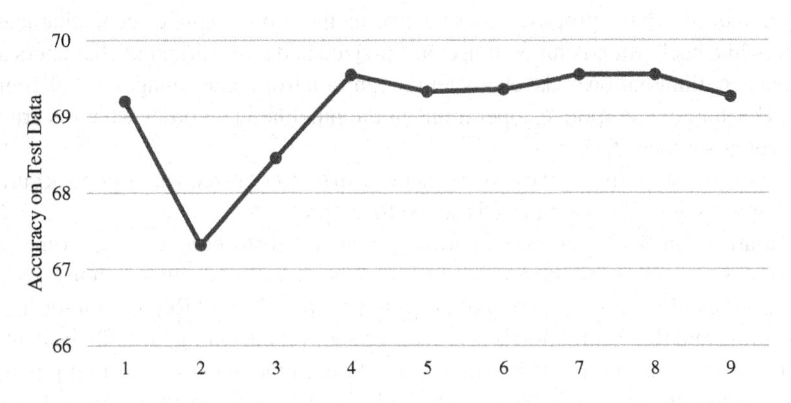


**Fig. 8** K-Nearest Neighbours with different values of k on HOG features

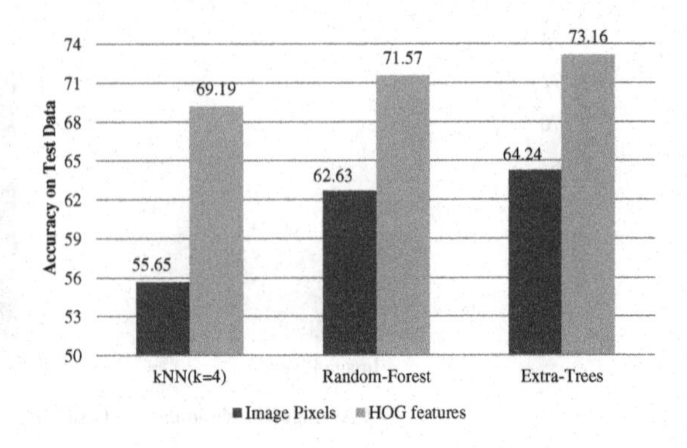


**Fig. 9** Comparison of accuracies of different classification techniques (on both feature inputs)

### *5.4 K-Nearest Neighbours Versus Random Forest Versus Extra Tree*

Ensemble methods outperform K-Nearest Neighbours significantly. Extra tree classifiers usually have trees with greater height than random forest. This implies that extra tree can generalize better but is computationally more expensive in comparison. In our case, extra tree classifier gives a noteworthy improvement over random forest classifier (Fig. 9).

## 6 Conclusion and Future Prospects

In summary, we have proposed an effective method to recognize scene characters. Factors like background clutter, noise, multi-sized and multi-oriented characters and variance in illumination make character recognition from scene images a challenging task. Our model incorporates operations in the pipeline in an order which is crucial to account for these factors.

Our results show that scene text characters can be recognized with a good accuracy in natural images, although there is scope for improvement.

Variations in thickness, illumination, geometric distortions arising from camera angles as well as writing variations, occlusion and non-random noise reduce effectiveness of feature extraction of proposed method. Kernel Regression for Image Processing and Reconstruction is less investigated, yet intriguing idea. Shape context can be possibly used for better feature extraction. In the machine learning pipeline, boosting and dimensionality reduction can be applied to improve accuracy. Convolutional neural networks are also a promising alternative.

This work can further help in building vision systems, which can solve higher level semantic tasks, such as word recognition and scene interpretation.

## References

1. Fraz, M., Sarfraz, M., Edirisinghe, E.: Exploiting colour information for better scene text recognition. In: Proceedings of the British Machine Vision Conference. BMVA Press (2014)
2. de Campos, T.E., Babu, B.R., Varma, M.: Character recognition in natural images. In: Proceedings of the International Conference on Computer Vision Theory and Applications, Lisbon, Portugal (2009)
3. Sheshadri, K., Divvala, S.K.: Exemplar driven character recognition in the wild. In: BMVC, pp. 1–10 (2012)
4. Dalal, N., Triggs, B.: Histograms of oriented gradients for human detection. In: IEEE Computer Society Conference on Computer Vision and Pattern Recognition, CVPR 2005, vol. 1, pp. 886–893. IEEE (2005)
5. Zhang, Z., Sturgess, P., Sengupta, S., Crook, N., Torr, P.: Efficient discriminative learning of parametric nearest neighbor classifiers. (2012)
6. Kumar, D., Ramakrishnan, A.: Recognition of kannada characters extracted from scene images. In: Proceeding of the Workshop on Document Analysis and Recognition, pp. 15–21. ACM (2012)
7. Neumann, L., Matas, J.: A method for text localization and recognition in real-world images. In: Computer Vision-ACCV 2010, pp. 770–783. Springer (2011)
8. deCampos, T.: The Chars74k Dataset. http://www.ee.surrey.ac.uk/CVSSP/demos/chars74k/ (2009). Accessed 5 Dec 2014]
9. 8.3, M., Release, C.V.S.T., 2014a, I.P.T.R.: version 8.3 (R2014a). The MathWorks Inc., Natick, Massachusetts (2014)
10. Pedregosa, F., Varoquaux, G., Gramfort, A., Michel, V., Thirion, B., Grisel, O., Blondel, M., Prettenhofer, P., Weiss, R., Dubourg, V., et al.: Scikit-learn: machine learning in python. J. Mach. Learn. Res. **12**, 2825–2830 (2011)
11. Otsu, N.: A threshold selection method from gray-level histograms. Automatica **11**, 23–27 (1975)
12. Gonzalez, R.C., Woods, R.E.: Morphological image processing. Digital Image Processing. Addison-Wesley Longman Publishing Co., Inc., Boston (2001)
13. Gonzalez, R.C., Woods, R.E.: Order-statistic (nonlinear) filters, example 3.14. In: Digital Image Processing, pp. 157. Pearson Education (2008)
14. Lim, J.S.: Two-Dimensional Signal and Image Processing, vol. 710, p. 1. Prentice Hall, Englewood Cliffs (1990)
15. Cover, T., Hart, P.: Nearest neighbor pattern classification. IEEE Trans. Inf. Theory **13**, 21–27 (1967)
16. Scikit-learn: sklearn.neighbors.KNeighborsClassifier Scikit-learn. (http://scikit-learn.org/stable/modules/generated/sklearn.neighbors.KNeighborsClassifier.html). Accessed 5 Nov 2014
17. Breiman, L.: Random forests. Mach. Learn. **45**, 5–32 (2001)
18. Scikit-learn: sklearn.ensemble.RandomForestClassifier. (http://scikit-learn.org/stable/modules/generated/sklearn.ensemble.RandomForestClassifier.html). Accessed 5 Nov 2014
19. Scikit-learn: Grid Search: Searching for estimator parameters. (http://scikit-learn.org/stable/modules/grid_search.html). Accessed 5 Nov 2014
20. Geurts, P., Ernst, D., Wehenkel, L.: Extremely randomized trees. Mach. Learn. **63**, 3–42 (2006)
21. Scikit-learn: sklearn.ensemble.ExtraTreesClassifier - Scikit-learn. (http://scikit-learn.org/stable/modules/generated/sklearn.tree.ExtraTreeClassifier.html). Accessed 5 Nov 2014
22. Shi, C.Z., Wang, C.H., Xiao, B.H., Zhang, Y., Gao, S.: Multi-scale graph-matching based kernel for character recognition from natural scenes. Acta Autom. Sinica **40**, 751–756 (2014)
23. Newell, A.J., Griffin, L.D.: Natural image character recognition using oriented basic image features. In: Proceedings of the 2011 International Conference on Digital Image Computing Techniques and Applications (DICTA), pp. 191–196. IEEE (2011)

# Adaptive Skin Color Model to Improve Video Face Detection

**Shrey Jairath, Samarth Bharadwaj, Mayank Vatsa and Richa Singh**

**Abstract** Face detection is a challenging problem having a wide range of security and surveillance-based applications. Skin color can be an important differentiator and used to augment the performance of automatic face detection. However, obtaining skin color consistency across illumination, different camera settings, and diverse ethnicity is a challenging task. Static skin color models that rely on image preprocessing are able to bring only limited consistency. Their performance in terms of accuracy and computation time degrades severely in real-world videos. In this paper, we study the dynamics of different color models on a database of five videos containing more than 93,000 manually annotated face images. Further, we propose an adaptive skin color model to reduce the false accept cases of Adaboost face detector. Since the face color distribution model is regularly updated using previous Adaboost responses, we find the system to be more effective to real-world environmental covariates. Importantly, the adaptive nature of the skin classifier does not significantly affect the computation time.

## 1 Introduction

*Accurate* face detection is a major challenge in real-time face recognition applications. Since recognition is a computationally expensive task, the onus is on the face segmentation module to provide detected faces from video frames in real time, with high confidence and low false accepts. In video face detection applications,

S. Jairath · S. Bharadwaj · M. Vatsa (✉) · R. Singh
Indraprastha Institute of Information Technology, New Delhi, India
e-mail: mayank@iiitd.ac.in

S. Jairath
e-mail: shrey08049@iiitd.ac.in

S. Bharadwaj
e-mail: samarthb@iiitd.ac.in

R. Singh
e-mail: rsingh@iiitd.ac.in



R. Singh et al. (eds.), *Machine Intelligence and Signal Processing*,
Advances in Intelligent Systems and Computing 390,
DOI 10.1007/978-81-322-2625-3_12

face detection algorithms with low false accept rate are more desirable since videos present plenty of opportunities to capture a face image. Adaboost face detector [7] has shown enduring performance in several applications. However, there is an important trade-off between face detection performance and false accepts of Adaboost. We have observed that the Adaboost face detector's overall performance can be significantly improved by *separately* pruning the number of false accepts. It is our hypothesis that adaptive (rather than static) skin color models, which are computationally cheap yet powerful classifiers can enhance the detection performance while maintaining low computation time.

## *1.1 Literature Review*

Skin color models are studied to a considerable extent in literature. It is well established that the static color models are excellent cues for applications such as image content understanding and content aware compression [6]. Moreover, Kakumamu et al. [2] have presented a comprehensive review of skin-based face detection algorithms and analyzed the performance of following skin models:

- Basic color models (RGB, normalized RGB)
- Perceptual color models (HIS, HSV)
- Perceptual uniform color models (CIE-Lab, CIE-Luv)
- Orthogonal color models (YCbCr, YIQ)

It is observed that YCbCr color model, particularly Cb and Cr channels, is most suitable in skin detection. This is because it is an orthogonal color space with reduced redundancy among channels, and lower statistical correlation among them.

Yang and Ahuja [9] used a single Gaussian model to determine Mahalanobis distance between color pixels and static skin thresholds. Other advanced techniques used in the literature to learn skin and non-skin labels are Naive Bayes classifier, variants of artificial neural networks and maximum entropy classifiers [2, 10]. However, these techniques are computationally expensive, require large training images and usually require image preprocessing to highlight color differences. Since, skin threshold does not necessarily encompass the appropriate color space, skin models alone for face detection leads to large false positives.

In video face detection and surveillance applications, particularly for videos that are affected by significant illumination variance, different camera settings and large ethnic bias, face detection is a major challenge. In recent literature, adaptive skin models have shown more promise in such challenging environments. Soriano et al. [4] propose a histogram back projection technique to trace the locus of the skin color in the color space in different camera settings. This technique has shown promising results to track face in web-camera feeds. Stern et al. [5] swap between different uncorrelated color spaces to improve skin detection in tracking. Despite some promise, the use of skin models alone is not a reliable solution to video face detection in surveillance applications.

Viola and Jones [7] proposed Adaboost face detection algorithm that is based on classifying Haar like features using a cascade of weak classifiers. This technique is found to be robust in various real-world applications of face detection. Further, several enhancements using skin models have been proposed. Wu et al. [1] use static parameterized skin model to create a candidate set of faces to improve Adaboost face detection. While the reduced search space had improved computation time, the fixed skin color threshold failed to detect skin in many occasions. Similarly, Ho et al. [8] use static heuristic thresholds as skin classifiers. Experimentally, we observed that such approaches do not substantially improve performance in challenging environments. This is predominantly due to the limitations of static color models. Also, in both [1, 8], results are showcased on a generated database that is not available to the research community.

### *1.2 Research Contributions*

In this research, a novel adaptive framework is proposed that integrates skin color classification with Adaboost to improve the face detection performance. The major contributions of this research work can be summarized as follows:

- A framework to utilize an adaptive skin classifier in real time to prune out the false accepts of Adaboost face detector is presented. We observe that the pruning is more effective than using skin models as a weak classifier in the Adaboost face detector. Further, the adaptive nature of the skin model and a fast update process provides more versatility to the system. Depending on the exact application, a large percentage of false accepts can be pruned at a small cost of reduction in genuine accepts. Details of the algorithm are presented in Sect. 2 and experiments are discussed in Sect. 3.
- This research also provides a database of videos containing over 93,000 manually labeled frames/images for face detection. These videos range from normal to very challenging scenarios. A detailed description of the database is provided in Sect. 3.1 which will be made public for research purposes.

## 2 Adaptive Skin Model for Improving Adaboost Face Detection

Previous research has focused on enhancing Adaboost face detector by combining it with different static skin color models. However, as mentioned previously, we have observed that existing static models do not necessarily improve Adaboost face detection results. The proposed algorithm (that is, tailored toward video surveillance applications), under Gaussian assumption, frequently updates the color model for wider reception to current environmental conditions. The proposed algorithm is shown in Fig. 1 and explained below.

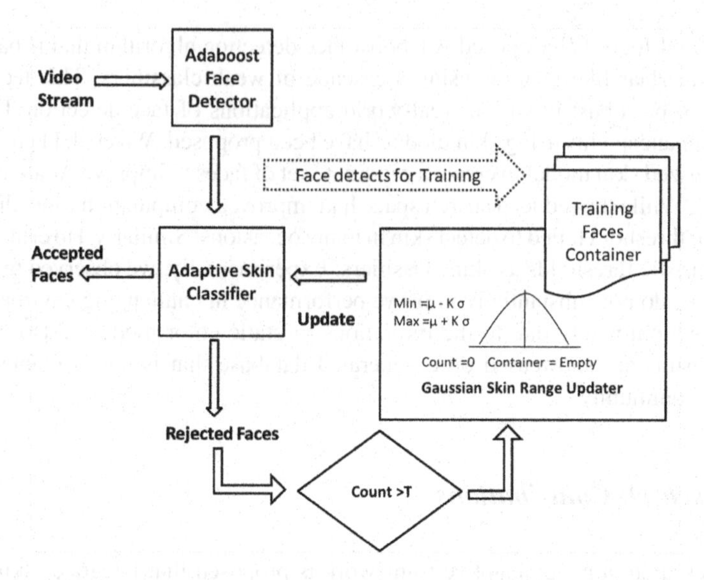


**Fig. 1** Block diagram of the proposed adaptive face detection approach

## *2.1 Skin Color Model as Gaussian Distribution*

Using the manually annotated face regions, we plot color histograms of Cb and Cr color channels of only the face regions. Figures 2, and 3 illustrates the histograms of these channels across all the videos in the database. While we concede that this data does not pass the Lilliefors test at 5 % significance, the color distribution can be approximated to Gaussian based on the bell shape of the histograms.

As discussed previously, under some specified illumination and environmental conditions, skin color occupies a certain region in the color space. Using, a relatively small and representative training set, this region can be modeled through a Gaussian probability distribution given by

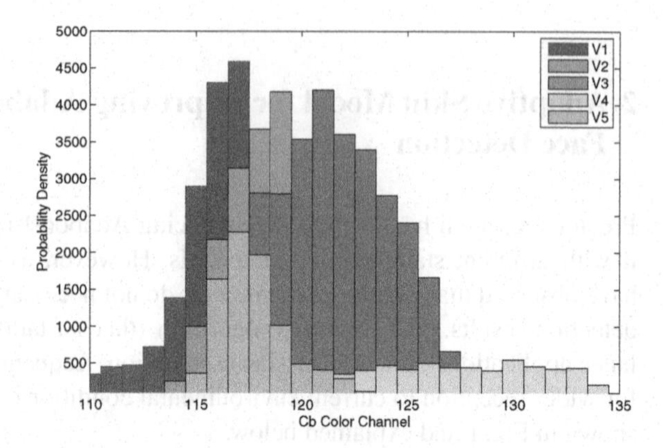


**Fig. 2** Histogram of Cb color channel pertaining to face regions in all the videos

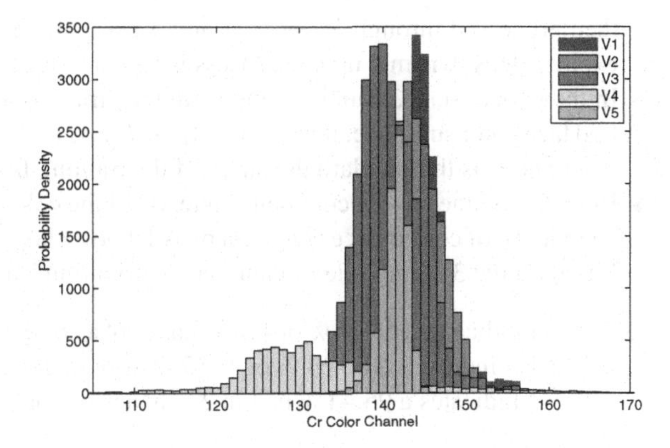


**Fig. 3** Histogram of Cr color channel pertaining to face regions in all the videos

$$P(x) = \frac{1}{\sigma\sqrt{2\pi}} e^{-(x-\mu)^2/2\sigma^2} \tag{1}$$

where $\mu$ is the mean skin color in the training set and $\sigma$ is the standard deviation. The experiments suggest that an adaptive skin model using Cb or Cr from YCbCr color model yields better results compared to RGB and HSV color channels. Furthermore, a color channel modeled through a single Gaussian is fast and requires small storage space. The model can be therefore quickly retrained using a small training set with negligible time delay.

## *2.2 Proposed Approach for Face Detection*

The proposed method described in Fig. 1 is as follows:

- Step 1: Adaboost face detector is first used to detect faces in the video frame and multiple faces may be detected. Adaboost may also detect regions that are false accepts.
- Step 2: To prune falsely accepted face detects, we use the $Cb$, $Cr$ color model. The mean color of each bounding box, that Adaboost has classified as face, is computed. The Adaboost label is said to be correct if the computed mean lies within a given confidence interval of the color model.
- Step 3: Mean of the bounding box of every Adaboost face is stored in the "training face container."
- Steps 1, 2, and 3 are repeated for subsequent video frames until the number of rejected Adaboost faces exceeds a tolerance $T$.
- *Tolerance* ($T$) is defined as the number of detected faces by Adaboost that the skin color model can reject before it is updated.
- After updating, the rejected candidate count is re-initialized to zero and the training face container is reset to empty for the new skin model. Subsequent frames are

then processed through Steps 1, 2, and 3 using the new skin model until it also expires. This dynamic update process is repeated for all the video frames.

- In the update stage, training samples in the "training face container" are used to find face color subspace. It is given as $[\mu - k * \sigma, \mu + k * \sigma]$, where $\mu$ is the mean color and $\sigma$ is the standard deviation of the training face container.
- Here, $k$ is defined as the confidence interval of the color model. This represents the confidence of correct face detection by Adaboost classifier for a given application. Based on the 3-sigma rule of Gaussian distribution [3], $k$ can have three values.

  - $k = 1$ indicates a 68.26 % or 1 in 3 chance of a correct classification by Adaboost
  - $k = 1.5$ indicates a 86.64 % or 1 in 7 chance of correct classification and
  - $k = 2$ indicates a 95.44 % or 1 in 22 chance of correct classification

- This approach creates dynamic color model of only the detected face rather than the entire color space. The values of $T$ and $K$ are provided based on the confidence on Adaboost classifier and the tolerance that characterizes a particular application.

For better understanding and reproducibility, we present the pseudo code of the algorithm as follows:

```
Final_Faces := null
Rejected_Candidate_Faces :=0
Training_Faces_Container := null
lower_X := 0
upper_X := 255

Proc Face_Detection (Video,T,k)
{
    while(Video_Frame in Video) {
    Candidate_Faces := AdaboostFD(Videoframe)
    for( Face in Candidate_Faces) {
    if(mean(Face) > lower_X && mean(Face)< upper_X)
        Final_Faces->Push(Face)
    else
        Rejected_candidate_Faces ++

    Training_Faces_Container->Push(mean(Face))
    if(Rejected_candidate_Faces > T)
        Update(k)
    }}
}

Proc Update(k) {
mean_skin = mean(Training_Faces_Container)
stddev_skin = Stddev(Training_Faces_Container)
lower_X = mean_skin - k * stddev_skin
upper_X = mean_skin + k * stddev_skin
}
```

**Table 1** Details of the annotated video database used for evaluation

| Video ID | Video description | Duration (in minutes) | Number of faces |
|---|---|---|---|
| V1 | Multiple faces in a frame, wide range of illumination levels and backgrounds | 21 minutes each | 22193 |
| V2 | | | 31505 |
| V3 | | | 23834 |
| V4 | High illumination variation | 12 | 7698 |
| V5 | Constant face with background variation | 15 | 8875 |

# 3 Experiments and Results

This section describes the database used, experiments, and the inferences drawn.

## *3.1 Annotated Video Database*

To the best of our knowledge, there is no publicly available annotated surveillance video database with multiple subjects in different illumination conditions. Therefore, an annotated video database of five videos (termed as V1, V2, V3, V4, and V5) each with 24 frames/second is prepared. Three videos V1, V2, and V3 are of television shows while videos V4 and V5 are prepared from movie clips. Detail characteristics of these videos are described in Table 1. Using an annotation tool, the videos are manually annotated (frame by frame) for the ground truth of true face locations. The videos are chosen such that the proposed algorithm can be evaluated with illumination, ethnicity, and background variations.[1]

## *3.2 Experimental Results*

Performance of the proposed algorithm is evaluated with respect to the improvement it makes to Adaboost face detector.[2] We first compute the performance when static/fixed RGB, HSV, and YCbCr models are used with Adaboost, similar to the

[1]We will make the annotated database, an OpenCV implementation of the annotation tool, and proposed algorithm publicly available for the researchers. To maintain anonymity, we have not included download details.

[2]The experiments were conducted on Pentium Core 2 Duo, 2.66 GHz with 4 GB RAM in real time.

**Table 2** Results on the static CbCr skin color models (results are reported as percentage reduction in GA/percentage reduction in false accept)

| Video ID | Results |
|---|---|
| V1 | 0.9/1.0 |
| V2 | 0.7/1.0 |
| V3 | 0.9/3 |
| V4 | 5.1/50 |
| V5 | 0/1.1 |

proposed approach. We also evaluate various combinations and observed CbCr as the best static model. However, these results suggest that static approaches are only marginally improving the performance as described in Table 2. In case of $CbCr$ color model, Table 2 shows the percentage decrease in Genuine Accept (GA) corresponding to the percentage decrease in the number of False Accepts (FA) of the overall face detection system. We next compute the results of the proposed adaptive algorithm. For this experiment, the results are reported for both $Cb$ and $Cr$ color channels across $k = [1, 1.5, 2]$ and $T = [1, 2 \ldots 10]$. Figure 4 summarizes the results of this experiment in tabular form whereas Fig. 5 shows sample frames with detection output.

### 3.3 Interpretation of Results

From the OpenCV implementation of the algorithm, it can be visually inferred that the overall performance of Adaboost face detector has been considerably improved. This can also be verified by viewing some visual examples in Fig. 4. In the interest of interpretation and proper analysis, we compared the output of the proposed algorithm with those obtained from Adaboost alone in the following manner.

- The number of genuine acceptance (GA) and false acceptance (FA) are computed for each video using Adaboost alone. This process is repeated for the proposed framework for a range of confidence (K) and tolerance (T).
- The trade-off is the percentage decrease in false accepts (FA) for a percentage decrease in genuine accepts (GA). The *best* values are dependent entirely on the practitioners and the specific application.

The trade-off of accuracy and false accepts in Adaboost is governed by it's parameters, namely, scale factor, initial window size, and minimum neighboring windows. It should be noted that in using Adaboost for the proposed framework, it is not necessary that competing performance be achieved with these same parameters. Hence, in Fig. 4, the results are represented in tabular format. Further, the highest decrease

| K | Video ID | | Tolerance | | | | | | | | | |
|---|---|---|---|---|---|---|---|---|---|---|---|---|
| | | | 1 | 2 | 3 | 4 | 5 | 6 | 7 | 8 | 9 | 10 |
| 1 | V1 | Cb | 23.8/38.0 | 20.1/37.3 | 19.5/38.6 | 18.0/37.2 | 17.2/35.1 | 21.0/38.3 | 21.0/44.6 | 18.0/38.23 | 20.5/41.2 | 20.4/42.7 |
| | | Cr | 23.1/40.0 | 20.2/40.0 | 21.9/39.5 | 20.4/39.5 | 19.7/40.2 | 21.0/42.2 | 21.0/40.6 | 22.2/42.5 | 20.6/42.2 | 22.7/40.1 |
| | V2 | Cb | 23.3/37.1 | 21.0/38.0 | 16.1/34.1 | 17.0/35.0 | 17.3/37.3 | 17.6/37.0 | 18.4/37.7 | 17.4/36.2 | 20.1/42.7 | 18.3/38.1 |
| | | Cr | 25.2/43.9 | 22.6/40.6 | 21.1/42.1 | 18.7/38.0 | 20.0/39.0 | 21.1/44.0 | 20.4/41.3 | 21.3/42.1 | 22.3/42.5 | 20.7/41.7 |
| | V3 | Cb | 25.1/53.1 | 17.1/25.7 | 21.2/49.5 | 21.0/45.5 | 17.3/23.8 | 15.2/28.7 | 17.7/42.0 | 17.6/48.2 | 19.0/41.5 | 20.0/46.0 |
| | | Cr | 27.0/50.1 | 23.9/46.9 | 24.0/48.0 | 24.0/51.0 | 19.1/44.0 | 20.8/47.9 | 18.4/43.1 | 20.2/45.3 | 20.3/44.4 | 20.4/47.9 |
| | V4 | Cb | 37.0/47.0 | 21.7/39.1 | 31.8/46.0 | 29.6/47.1 | 27.2/44.4 | 27.5/46.3 | 21.0/41.2 | 24.4/46.4 | 23.4/47.6 | 22.7/50.0 |
| | | Cr | 47.9/54.6 | 38.8/51.6 | 32.8/47.3 | 32.2/50.7 | 30.1/47.5 | 28.8/49.8 | 31.1/52.3 | 28.2/55.0 | 25.2/50.7 | 22.4/50.0 |
| | V5 | Cb | 19.6/29.7 | 13.8/23.5 | 14.1/25.6 | 10.5/22.7 | 9.8/21.2 | 8.4/20.0 | 8.1/19.3 | 10.0/24.6 | 9.6/24.0 | 5.5/20.5 |
| | | Cr | 17.0/32.6 | 16.0/29.5 | 11.7/26.0 | 13.0/28.7 | 12.8/30.8 | 12.1/26.7 | 11.7/28.0 | 14.5/31.0 | 14.5/32.0 | 15.0/31.0 |
| 1.5 | V1 | Cb | 4.4/11.7 | 3.1/8.7 | 3.1/10.6 | 4.1/11.6 | 3.7/11.6 | 3.4/11.8 | 2.8/8.8 | 4.5/14.1 | 0.8/2.6 | 4.2/14.6 |
| | | Cr | 3.2/12.3 | 4.6/14.1 | 5.0/16.0 | 5.1/18.0 | 6.6/19.6 | 4.9/14.5 | 5.1/20.0 | 5.3/18.5 | 5.0/20.1 | 5.0/16.4 |
| | V2 | Cb | 3.4/10.8 | 3.8/11.7 | 3.0/9.6 | 3.4/10.3 | 0.4/3.1 | 5.7/17.6 | 1.0/4.4 | 4.7/12.8 | 5.5/15.0 | 5.0/15.4 |
| | | Cr | 4.0/10.0 | 4.1/12.0 | 4.7/14.6 | 4.2/12.2 | 5.0/16.3 | 6.2/18.8 | 3.8/16.3 | 4.7/18.5 | 5.1/22.0 | 4.7/19.5 |
| | V3 | Cb | 3.7/12.2 | 3.9/9.0 | 3.0/7.7 | 4.8/10.4 | 4.8/14.4 | 4.7/11.1 | 5.1/12.5 | 5.4/12.5 | 4.7/11.0 | 5.7/11.0 |
| | | Cr | 4.4/14.4 | 4.4/12.4 | 4.7/13.8 | 4.4/17.6 | 5.0/17.5 | 5.3/17.4 | 5.0/16.1 | 4.9/16.2 | 5.5/16.5 | 6.2/17.5 |
| | V4 | Cb | 12.8/18.0 | 15.5/27.7 | 8.9/20.0 | 9.6/24.5 | 7.8/19.4 | 11.6/31.2 | 10.2/25.3 | 10.3/28.0 | 10.6/29.1 | 10.0/28.3 |
| | | Cr | 24.8/31.6 | 11.1/24.0 | 15.3/26.5 | 12.3/30.3 | 9.4/29.1 | 10.0/28.6 | 17.0/33.4 | 13.2/34.0 | 10.8/30.6 | 10.7/30.2 |
| | V5 | Cb | 1.3/5.8 | 1.4/6.4 | 1.0/6.8 | 1.1/8.5 | 1.5/7.3 | 2.3/9.4 | 1.1/9.0 | 1.0/9.0 | 1.5/9.8 | 1.4/8.8 |
| | | Cr | 1.7/3.7 | 0.5/3.0 | 0.8/3.0 | 1.1/3.0 | 0/0.17 | 0/0.13 | 0/0.15 | 0/0.2 | 0/0.21 | 3.0/8.0 |
| 2 | V1 | Cb | 0.19/0.7 | 0.3/0.65 | 0.4/1.6 | 0.9/4.6 | 1.3/6.6 | 0.9/3.1 | 1.4/3.3 | 1.1/3.2 | 0.9/1.6 | 0.7/2.6 |
| | | Cr | 1.0/3.3 | 1.5/7.1 | 0.8/4.6 | 0.3/2.0 | 1.1/5.2 | 0.2/1.0 | 1.1/10.9 | 1.8/10.1 | 1.0/7.5 | 1.2/6.7 |
| | V2 | Cb | 0.1/0.2 | 0.04/0.2 | 0.6/4.0 | 0.1/0.4 | 1.8/8.3 | 0.1/0.4 | 0.16/0.6 | 0.16/0.8 | 0.2/0.7 | 1.0/6.1 |
| | | Cr | 0.3/2.2 | 0.6/2.6 | 0.4/3.7 | 0.7/5.0 | 1.1/5.7 | 1.5/7.9 | 0.5/4.9 | 0.7/5.1 | 0.7/5.0 | 1.7/9.1 |
| | V3 | Cb | 1.0/3.0 | 1.6/3.7 | 1.9/6.6 | 1.7/4.9 | 1.6/4.6 | 1.5/4.4 | 1.1/3.6 | 1.5/4.3 | 1.6/4.6 | 1.8/6.0 |
| | | Cr | 0.8/3.7 | 1.1/4.6 | 0.9/4.5 | 1.7/6.4 | 1.1/6.4 | 1.9/7.9 | 1.4/6.2 | 1.6/6.4 | 2.0/8.6 | 1.9/5.0 |
| | V4 | Cb | 5.7/9.8 | 3.2/7.0 | 2.9/10.1 | 2.2/3.8 | 3.4/8.7 | 4.8/11.5 | 2.1/7.3 | 4.5/7.5 | 3.1/7.3 | 3.1/8.2 |
| | | Cr | 5.8/10.5 | 5.8/11.1 | 1.5/5.0 | 3.2/8.1 | 3.0/9.6 | 5.6/17.7 | 3.7/14.1 | 7.14/18.1 | 3.2/14.7 | 5.7/15.2 |
| | V5 | Cb | 0.3/2.2 | 0.2/2.5 | 0.4/3.0 | 0.2/1.3 | 0.2/4.0 | 0.4/5.0 | 0.02/1.1 | 0.02/1.1 | 0.05/1.4 | 0.1/4.7 |
| | | Cr | 0.7/1.5 | 0/0.09 | 0/0.1 | 0.2/0.4 | 0/0.13 | 0/0.14 | 0/0.9 | 0.01/1.0 | 0/1.1 | 0.1/1.5 |

**Fig. 4** Results of the proposed adaptive algorithm. The results are presented as Percentage reduction in GA/Percentage reduction in FA

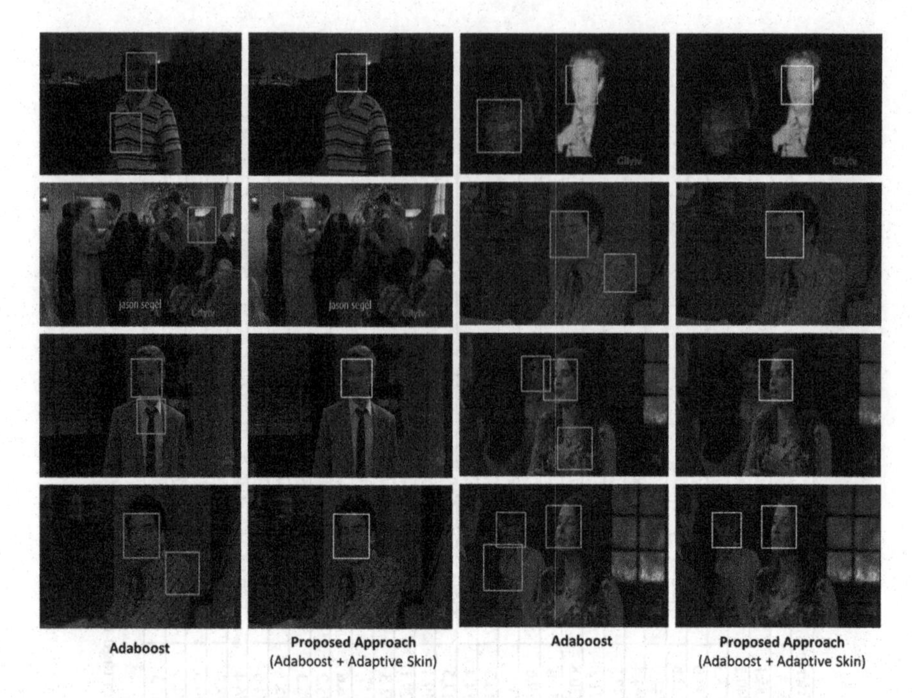


**Fig. 5** Results of the proposed and Adaboost face detectors. These results are computed at $T = 10$ and $k = 1.5$

in face rejects for each video are **emphasized.** It can also be inferred that since, in all cases, the percentage reduction in GA is lower than percentage reduction in FA, for all values of k and T, pruning has successfully taken place.

## *3.4 Analysis*

Key inferences of the results are as follows:

- At the confidence interval of $k = 1$, $V1$, $V2$, $V3$ (TV serial videos) show around 42 % decrease in false accepts for an average of 19 % loss in genuine accepts in comparison to Adaboost face detector.
- Similarly, at $k = 1.5$, $V1$, $V2$, $V3$, show a decrease of about 18 % false accepts for an average of 4.5 % decrease in genuine accepts.
- The proposed algorithm yields good performance even on challenging videos. For video $V5$ at $k = 1.5$ and $T = 8$, false accept reduces by 9 % for a decrease in genuine accept rate of 1 %.
- The exact values of parameters $k$ and $T$ for optimal performance of the system are dependent on the application, type, genre, and quality of the videos. In surveillance applications such as banks, the illumination variation is gradual and hence, $k$ can be larger.

- As $k$ is increased, the percentage reduction in Genuine accepts and False accepts decreases. Therefore, the value of $k$ is chosen based on the appetite for loss of genuine accepts and need for reduction of false accepts.
- Though some loss in genuine accept rates occur in comparison to Adaboost in reducing the false accepts, one must remember that this represents the number of face detects and not the actual subjects. The maximum tolerance value is 10 which is considerably less than the number of face detects that can occur in one second of a video.
- Since all videos in this experiment run at 24 frames per second and if a subject produces a detectable face for only half a second, it still implies 12 chances to get a good capture. Correspondingly, the reduction in false accepts will save computation time and unnecessary allocation of resources in real-time applications. In our experiment, we observe that the algorithm has detected each individual in the video but not necessarily all images of the same person. Therefore, the detection approach is effective for face recognition applications.
- In Fig. 5, first two columns show that the proposed algorithm reduces the false accepts obtained by Adaboost. In the third frame of last two columns, there is a false reject by the proposed algorithm, but this lead to updating skin model and subsequently the face is detected (fourth frame in last two columns).
- We also observe that the proposed algorithm provides better results than existing skin color-based approaches.

## 4 Conclusion

The paper presents an adaptive color model that learns from Adaboost face classifier to enhance the performance by pruning the false accepts. Unlike previous research, the proposed algorithm is invariant to illumination, background, and ethnicity bias since the face color model is continuously updated. This method is fast to implement and has several benefits in real-time video surveillance applications. In future, we plan to extend the algorithm by using advance machine learning approaches like active learning to update the color models.

## References

1. Hou, Y., Peng, Q.: Face detection based on adaboost and skin color. In: International Symposium on Information Science and Engineering, pp. 407–410 (2009)
2. Kakumanu, P., Makrogiannis, S., Bourbakis, N.: A survey of skin-color modeling and detection methods. Pattern Recognit. **40**(3), 1106–1122 (2007)
3. Preston, S.: The 68–95-99.7 rule for normal distributions. World Wide Web Electronic Publication (2010)

4. Soriano, M., Martinkauppi, B., Huovinen, S., Laaksonen, M.: Skin detection in video under changing illumination conditions. In: Proceedings of International Conference on Pattern Recognition, vol. 1, pp. 839–842 (2000)
5. Stern, H., Efros, B.: Adaptive color space switching for face tracking in multi-colored lighting environments. In: IEEE International Conference on Automatic Face and Gesture Recognition, pp. 249–254 (2002)
6. Vezhnevets, V., Sazonov, V., Andreeva, A.: A survey on pixel-based skin color detection techniques. In: Proceedings of the Graphicon, pp. 85–92 (2003)
7. Viola, P., Jones, M.J.: Robust real-time face detection. Int. J. Comput. Vis. **57**, 137–154 (2004)
8. Wu, Y.W., Ai, X.: An improvement of face detection using adaboost with color information. In: International Colloquium on Computing, Communication, Control, and Management, pp. 317–321 (2008)
9. Yang, M.H., Ahuja, N.: Detecting human faces in color images. In: IEEE International Conference on Image Processing, pp. 127–130 (1998)
10. Zhang, Z., Gunes, H., Piccardi, M.: An accurate algorithm for head detection based on xyz and hsv hair and skin color models. In: IEEE International Conference on Image Processing, pp. 1644–1647 (2008)

# Improving Rating Predictions by Baseline Estimation and Single Pass Low-Rank Approximation

**Shisagnee Banerjee and Angshul Majumdar**

**Abstract** In this work, we propose a single pass low-rank matrix approximation technique for collaborative filtering. The unknown values in the partially filled ratings' matrix is imputed by robust baseline prediction. The resulting matrix is not low-rank; but it is known from latent semantic analysis that the ratings matrix should be so since the number of factors guiding the users' choice of items is limited. Following this analysis, we compute a low-rank approximation of the filled ratings matrix. This is a simple technique that requires computing the SVD only once—unlike more sophisticated matrix completion techniques. We compared our proposed method with state-of-the-art matrix completion and matrix factorization-based collaborative filtering approaches and found that our proposed method yields significantly better results. The mean absolute error (MAE) from competing techniques is around 78 % where as ours is around 74 %.

**Keywords** Compressed sensing · EEG · WBAN

## 1 Introduction

Today recommender systems find abundant applications in Business-To-Client (B2C) eCommerce portals, e.g., Amazon, Netflix, eBay, Hulu, etc. give us personalized recommendations. The workhorse behind recommender systems is collaborative filtering. Collaborative filtering recommends items based on two types of data—explicit and implicit ratings.

Explicit ratings are sometimes collected by these portals, e.g., with questions like 'How I would rate the movie on a five point scale', 'If I would like or dislike a video

S. Banerjee · A. Majumdar (✉)
Indraprastha Institute of Information Technology – Delhi, New Delhi, India
e-mail: angshul@iiitd.ac.in

S. Banerjee
e-mail: shisagnee1423@iiitd.ac.in





R. Singh et al. (eds.), *Machine Intelligence and Signal Processing*,
Advances in Intelligent Systems and Computing 390,
DOI 10.1007/978-81-322-2625-3_13

(in Youtube)' or 'What would I rate a book', etc. The explicit ratings are harder to collect (since users are not always enthusiastic about providing this information), but are reliable.

The problem of explicit ratings is the sparsity—since users typically rate a handful of items (less than 10 %). This issue is ameliorated if instead of using such ratings, predictions are based on implicit data such as user's buying pattern or browsing history. Such information is called implicit ratings. Implicit ratings are easier to collect, but are not very reliable. For example, in a family computer, the father, the mother as well as the children may log in and look for different types of goods and services. If collaborative filtering tries to use such data of such diverse interests to predict products, the prediction is not likely to be correct for either of the users.

The term collaborative filtering means that the recommendation for a particular user is not only dependent on the past history of the user but also on the choices of other users on similar products/items. The basic problem in collaborative filtering is to predict the user's rating on an item. Assuming that the rating is correct, the system decides whether or not to recommend the item to the user (it recommends only if the predicted rating is high). There are two classes of algorithms falling under the purview of collaborative filtering.

In the initial days, collaborative filtering automated real-life scenarios. To predict the rating of a user on an item, the system searched the ratings database for similar users. Collaborative filtering interpolated the rating on the particular item by linearly weighting the ratings of the similar items on the item; users with higher similarities were given larger weights. Such an approach is called user-based recommendation [1]. The same can be done from the item's perspective. To find the user's rating on an item, similar items can be searched for and the unknown rating can be interpolated by weighing the ratings of the user on similar items. Such an approach is called item-based recommendation [2]. Both approaches fall under the category of neighborhood-based or memory-based systems.

These simple techniques were easy to interpret and analyze. The user's knew why the item was predicted; many a time the system mentioned the reason behind the recommendation, e.g., some TV show recommendation system explicitly mentions 'we are recommending you these shows because you rated highly some other show'. But these simple techniques had poor prediction accuracy. This was not good for the eCommerce portal. Bad recommendations mean that the user will not buy/rent the suggested items—this will reflect poorly on revenue. On the other hand, the user will also lose faith in the recommendation system and in the future may not follow its suggestions. Such negative vibe will also affect the portal adversely. Therefore, the primary interest of these portals is to improve recommendation accuracy. The Netflix prize is well known in this respect. They offered one million dollar prize to anyone who could reduce prediction error by 10 % compared to the benchmark (such memory based methods). This prize catapulted collaborative filtering to the popularity it enjoys today.

The second approach to collaborative filtering is based on the latent factor model [3]. To understand this model one needs to take a step back from collaborative filtering and be aware of the basic premises of content-based filtering. Content-based filtering tried modeling the user and the items under various topics. For example, for book recommendation these topics could be author, genre, subject, publisher, etc. The item (book) was characterized based on these topics—this was trivial. For the user, the same topics needed to be learnt from data (this was difficult) to build the user profile. Once the profiles were built, template matching between the user and the items were carried out to suggest recommendations (based on good match). Content-based filtering failed in recommendation systems; we do not want to discuss those reasons here.

Latent factor model abstracts content-based filtering. It only assumes that the user's rating on an item is guided by only a few factors. Latent factor model does not need to know these factors (unlike content-based filtering)–hence, the term 'latent.' It assumes that both the user and the item are characterized by the same factors. The items' possess these factors and the users have affinity toward these factors.

The users' ratings on items can be structured into a matrix with users along the rows and items along the columns. The ratings matrix is sparsely populated since users typically rate a handful of items. The problem is to predict the missing ratings. Latent factor models assume that the matrix is low-rank, this is because the number of factors guiding the ratings are very few—typically around 20 for movie recommendation even though the number of items and users may be in 1000 s or more. In latent factor models, the missing ratings are predicted by exploiting the low-rank nature of the matrix.

Usually the latent factors for the users and the items are computed from the available ratings. These are matrix factorization techniques. The problem with such an approach is that, it is a bi-linear nonconvex problem with no global convergence guarantees. To overcome this issue modern techniques propose to direct fill up the ratings matrix (without the need for estimating the latent factors). The ensuing problem is convex with known convergence and recovery guarantees.

Unfortunately, either of these techniques are time consuming since they require solving complex optimization problems. This is the main deterrent behind widespread application of these techniques in commercial recommender systems—the simple memory-based techniques are still preferred in practice. In this work, we propose a simple solution that does not require solving such difficult problems but still keeps the essence of latent factor modeling. Our solution is fast and only takes a fraction of the time required by the more sophisticated counterparts; but without degradation of prediction accuracy. In fact our method is significantly better than the more sophisticated counterparts both in terms of speed and accuracy.

In the following section, we review the relevant studies pertinent to our work. Section 3 describes our proposed approach. The experimental results are shown in Sect. 4. Finally in Sect. 5, the conclusions of the work are drawn.

## 2 Literature Review

### *2.1 Bias Prediction*

A rating consists of two parts, a baseline and an interaction.

$$r_{i,j} = b_{i,j} + z_{i,j} \tag{1}$$

Here, $i$ denotes the user and $j$ denotes the item; $b_{i,j}$ is the baseline consisting of the user's and the item's biases and $z_{i,j}$ is the interaction between the users and the items. The memory-based methods or the latent factor analysis are used to model the interaction. So far, we have not discussed about biases and baseline.

There are some users who are overtly critical and rate everything on the lower side of the scale. There are some items, for example, Booker winning books or Oscar winning movies—these are naturally hyped and enjoy a positive bias from all users. In the first case, the user's bias is negative and the second, the item's bias is positive. Such biases are independent of the interaction between the users and the items and are individualistic. Baseline is modeled using the biases as:

$$b_{i,j} = \mu + b_i + b_j \tag{2}$$

Here $\mu$ is the global mean. The user's bias $b_i$ is the deviation of the user from the mean, if it is negative the implication is that the user is critical and rates everything on the bottom half of the scale. Similarly, the item bias models the popularity of the product.

The biases need to be estimated from the data. There are two techniques or bias estimation. Koren and Bell popularized [4] the data fitting-based approach. They proposed a technique that estimated the biases such that the squared mismatch between the baseline and the ratings is minimized.

$$\min_{b_i,b_j} \left\| r_{i,j} - (\mu + b_i + b_j) \right\|_2^2 + \lambda \left( \|b_i\|_2^2 + \|b_j\|_2^2 \right) \tag{3}$$

The data mismatch term tries to minimize the error between the baseline and the rating. But it should be remembered that there is also the interaction (1). Since the interaction is not accounted for in (3), the regularization parameter $\lambda$ prevents over-fitting. Usually, stochastic gradient descent is used to estimate the user and item biases.

There is a simpler way to estimate the baselines. This is the Potter estimation. The formula is given by:

$$b_i = \frac{\sum\limits_{j \in R(i)} (r_{i,j} - \mu)}{\lambda + |R(i)|}, \; b_j = \frac{\sum\limits_{i \in R(j)} (r_{i,j} - \mu - b_i)}{\lambda + |R(j)|} \tag{4}$$

Here *R(i)* and *R(j)* denotes the neighborhood of the user and the item. |.| denotes the cardinality of the set.

The Potter estimation is a simple one shot method. It may seem that the first technique (3) is more mathematically appealing, but in practice both of them yield very similar results.

## 2.2 Latent Factor Model

The latent factor model characterizes the users and items by vectors of latent factors. For the $i$th user the latent factor vector is represented as:

$$u_i = [u_i^1, u_i^2, \ldots, u_i^f]^T \tag{5}$$

Similarly the item factor vector is represented as:

$$v_j = [v_i^1, v_i^2, \ldots, v_i^f]^T \tag{6}$$

When these factors match, the user likes the item and rates it high. This is simple to model, the interaction between the user $i$ and the item $j$ is expressed as an inner product between the user's and the item's latent factors.

$$z_{i,j} = u_i^T v_j \tag{7}$$

The model can be extended for the full ratings matrix. The user latent factor matrix constitutes the latent factors of all the users; this is expressed as:

$$U = \begin{bmatrix} u_1^T \\ \ldots \\ u_M^T \end{bmatrix} \tag{8}$$

Similarly the items' latent factor matrix is constructed as:

$$V = [v_1 \mid \ldots \mid v_N] \tag{9}$$

Therefore, the interaction matrix for all the users and the items is modeled as a product of the users' and the items' latent factor matrices.

$$Z = UV \tag{10}$$

Since the ratings matrix is not fully observed, the data model using the latent factor analysis is expressed as follows:

$$Y = R \odot Z = R \odot UV \tag{11}$$

where $R$ is the subsampling mask consisting of 1's at the positions where the ratings are observed and 0's elsewhere; $Y$ is the available interaction data (after bias correction).

Matrix factorization methods are used to recover $U$ and $V$. Usually, a regularized matrix factorization method is used.

$$\min_{U,V} \|Z - R \odot UV\|_F^2 + \lambda \left( \|U\|_F^2 + \|V\|_F^2 \right) \tag{12}$$

The problem of matrix factorization is that, it is a bi-linear problem and hence nonconvex. Consequently, there is no global convergence guarantee. Moreover, in practice one does not require estimating the latent factor matrices—the goal is predict the missing interactions. Both these issues can be solved simultaneously by moving away from the factorization approach to a completion approach.

The matrix $Z$ is low-rank—the rank is $f$ (number of factors) whatever be the number of users ($M$) and items ($N$). Under such conditions, the task is to find a low-rank approximation of $Z$. The straightforward solution is:

$$\min_{Z} rank(Z) \text{ such that } Y = R \odot Z \tag{13}$$

Unfortunately, rank minimization is an NP hard problem with doubly exponential complexity. Mathematicians have showed that it is possible to relax the NP hard norm minimization problem to its nearest convex surrogate—the nuclear norm, and still be able to get results equivalent to (13). Nuclear norm minimization is expressed as,

$$\min_{Z} \|Z\|_{NN} \text{ such that } Y = R \odot Z \tag{14}$$

The nuclear norm is defined as the sum of singular values.

The aforesaid problem (14) is convex and can be solved via semi-definite programming. The solution has convergence guarantees, and the recovery guarantees via (14) has been proven [5].

Unfortunately, solving (14) is computationally challenging. It requires the computing the SVD in every iteration. The complexity of SVD is O($n^3$). Solving the matrix factorization problems, although cheaper than nuclear norm minimization are time consuming nonetheless. In this work, we propose a technique to solve the collaborative filtering problem in a computationally cheap fashion.

## 3 Proposed Approach

A simple and efficient approach to solve the collaborative filtering problem was proposed in [6]. In the first step, it imputes missing values in the ratings matrix by the corresponding column averages. Then, the row averages are subtracted to get a centered matrix. From latent factor models, it is known that the ratings matrix should be low-rank. But, after such missing value imputations, the resulting matrix is not so. Therefore in [6], a low-rank approximation of the resulting matrix is computed. This is done by computing the SVD and keeping the top 'f' singular values and corresponding singular vectors.

This is a naïve approach. The column averages do not represent the ratings in any fashion. In this work, we propose to improve upon it by imputing the missing values not from arbitrary column / row averages but by robust baseline prediction. The steps of our proposed approach are the following:

1. Impute the missing values in the ratings matrix $R$ by baseline prediction. The baseline is predicted using (2). The user and item biases are computed using simple Potter estimation (4).
2. Compute the SVD of the filled matrix (say $A$): $A = USV^T$
3. Compute a low-rank approximation of $R$ by keeping the top 'f' singular values and the corresponding singular vectors: $\hat{R} = U(1:\mathrm{f})S(1:\mathrm{f})\mathrm{V}(1:\mathrm{f})^T$

The cost of computing the SVD is $O(n^3)$. In our proposed approach we need to compute the SVD only once. The cost of computing the biases is only $O(n)$ and is negligibly small compared to the cost of computing the SVD. Unlike matrix factorization methods our technique does not require solving least squares problems in every iteration. The cost of solving a least squares problem is $O(n^3)$; it needs to be solved twice every iteration for many iterations. For matrix completion-based techniques one needs to compute SVD is every iteration; These techniques also need many iterations to converge.

# 4 Experimental Evaluation

Experiments were carried out on a standard movie ratings database. MovieLens data sets were collected by the GroupLens Research Project at the University of Minnesota. The 100 K data set consists of 100,000 ratings (1–5) from 943 users and 1682 items. Each user has rated at least 20 movies. The data was collected through the MovieLens web site(movielens.umn.edu) during the 7-month period from September 19th, 1997 through April 22nd, 1998. This data has been cleaned up—users who had less than 20 ratings were removed from this data set.

We compare our technique with state-of-the-art techniques in matrix completion (MC) [7] and matrix factorization (MF) [8]. The split Bregman technique for matrix completion [7] has been empirically shown to outperform other matrix completion techniques for general matrix completion problems. The bayesian matrix factorization (BMF) method [8] has shown to yield very good results for collaborative filtering. The MC and the MF techniques are directly applied on the ratings matrix without bias correction.

**Table 1** Comparison of MAE

| # factors (f) | MC [7] | BMF [8] | Proposed | SVD-CF [6] |
|---|---|---|---|---|
| 20 | 0.7803 | 0.7803 | **0.7369** | 0.7911 |
| 30 | 0.7803 | 0.7785 | **0.7387** | 0.7934 |
| 40 | 0.7803 | 0.7769 | **0.7409** | 0.7922 |
| 50 | 0.7803 | 0.7798 | **0.7424** | 0.7941 |

In the following table, we show the mean absolute error (MAE) from our proposed approach vis-à-vis MC and MF for varying number of factors (not applicable for MC). We also compare the results with the naïve SVD-CF method [6] (Table 1).

We observe that our proposed simple approach triumphs over its more sophisticated counterparts. The results are significantly better. In the spirit of reproducible research we have made our implementation available at [9].

Although experiments were carried out on a Intel Core i7 with 4 GB of RAM running on Windows 7. The matrix completion technique takes 101 s to run; the run-times for the BMF is 135 s. The run-time for our proposed approach (including baseline prediction) is 5 s. The SVD-CF technique is slightly faster as it does not require computing biases; it takes 3 s on an average. Our algorithm is more than an order of magnitude faster than MC and BMF.

## 5 Conclusion

In this work, we propose a simple technique for rating prediction. Our approach is grounded in the latent factor model. But instead of using computationally complex low-rank matrix factorization and matrix completion techniques, we propose a single pass low-rank approximation method with baseline prediction. The technique yields significantly better results (in terms of MAE) over its more sophisticated counterparts. We reduce the error by around 4 %. Moreover, our method is more than an order of magnitude faster compared to these techniques. This is because, ours is a single pass method which requires computing the SVD only once.

## References

1. Herlocker, J.L., Konstan, J. A., Borchers, A., Riedl, J.: An algorithmic framework for performing collaborative filtering. In: ACM SIGIR, pp. 230–237 (1999)
2. Sarwar, B., Karypis, G., Konstan, J., Riedl, J.: Item-based collaborative filtering recommendation algorithms. WWW 285–295 (2001)
3. Hofmann, T.: Latent semantic models for collaborative filtering. ACM Trans. Inf. Syst. **22**, 89–115 (2004)

4. Koren, Y., Bell, R.: Recommender systems handbook. Advances in Collaborative Filtering, pp 145–186. Springer, New York (2011)
5. Candès, E.J., Plan, Y.: Matrix completion with noise. Proc. IEEE **98**(6), 925–936 (2009)
6. Vozalis, M., Markos, A., Margaritis, K.: Evaluation of standard SVD-based techniques for Collaborative Filtering. In: Proceedings of the 9th Hellenic European Research on Computer Mathematics and its Applications (2009)
7. Gogna, A., Shukla, A., Majumdar, A.: Matrix recovery using split bregman. In: ICPR (2014)
8. Yuan, C.: Multi-task learning for bayesian matrix factorization. In: IEEE ICDM (2011)
9. http://www.mathworks.in/matlabcentral/fileexchange/48240-rank-approximation

# Retinal Blood Vessel Extraction and Optic Disc Removal Using Curvelet Transform and Morphological Operation

**Sudeshna Sil Kar and Santi P. Maity**

**Abstract** This paper proposes an algorithm for automatic blood vessel extraction and optic disc removal on retinal images using curvelet transform and morphological operation. Since the curvelet transform represents the lines, the edges, the curvatures, the missing and the imprecise boundary details compactly, i.e., by smaller number of coefficients, they can be tuned suitably to enhance the image details. To remove the optic disc, the edge enhanced image is first opened by a disc-shaped structuring element which is then subtracted from the inverted histogram equalized image. Next, the poor dynamic range of the subtracted image is enhanced followed by multi-structure opening with linear structuring element to extract the vessel structures and remove the spurious components. Extensive simulations on publicly available DRIVE database show that the present work outperforms the existing works for various types of vessels extraction and the optic disc removal.

**Keywords** Retinal image segmentation · Vessel detection · Curvelet transform · Morphology

## 1 Introduction

Automatic and accurate extraction of retinal blood vessels play an important role for the diagnosis of different ophthalmologic diseases including glaucoma, diabetic retinopathy (DR), etc. [10]. For example, the abnormal narrowing of retinal blood vessels may indicate the earlier stages of glaucoma. DR is characterized by the growth of new tiny, thin, fragile, and abnormal vascular nets (neovascularization) which cause frequent minor bleeding that may lead to permanent vision loss also. The poor

S.S. Kar (✉) · S.P. Maity
Department of Information Technology, Indian Institute of Engineering Science and Technology, Shibpur, Howrah 711103, West Bengal, India
e-mail: sudeshna.sil@gmail.com

S.P. Maity
e-mail: santipmaity@it.iiests.ac.in



R. Singh et al. (eds.), *Machine Intelligence and Signal Processing*, Advances in Intelligent Systems and Computing 390,
DOI 10.1007/978-81-322-2625-3_14

contrast between the background and the vessels, the disconnected and unsharp vessel boundaries make most of the narrow vessels almost inseparable from the background. Since the vessels emerge from the optic disc (OD), to extract the complete vessel structure particularly those vessels that lie in the OD region, suppression of the OD is very much essential prior to vessel segmentation. Otherwise, it may hamper the further medical investigation process.

For vessel extraction, the methods available in the literature may broadly be classified as pattern recognition techniques [5], matched filtering [2], vessel tracking/tracing [1], mathematical morphology [3], etc. A detailed description of these methods can be found in [4]. Among them, matched filtering using Gaussian kernel is widely used due to its optimal performance in detecting vessels which also have Gaussian cross-sectional profile. However, matched filtering not only extracts the vessels accurately but also detects the OD boundary due to its high matched filter response which in turn increases the false detection rate. Suppression of the OD is therefore necessary prior to matched filtering which has been addressed in the literature by morphological operation. Morphological operations on the other hand can remove the OD efficiently and extract the thick and the long vessels accurately but fails to extract the tiny and the weak vessels since they are inseparable from the background. The extraction of the thin, the weak and the faint vessels using morphology needs some sort of preprocessing operation in the form of contrast/edge enhancement prior to vessel extraction.

In this paper, curvelet transform is used for edge enhancement since it is very much efficient in identifying curves, contours, missing and imprecise edges, curvatures, and other boundary information. Moreover, being a member of wavelet family, it offers multiresolution, space-frequency localization ability, very high directional sensitivity and anisotropy in frequency domain. The edge enhanced image is then opened by a disc-shaped structuring element (SE) and is subtracted from the inverted histogram equalized image. Next, the poor dynamic range of the subtracted image is enhanced followed by multi-structure opening with linear SE to extract the vessel structures and remove the spurious components.

The rest of the paper is organized as follows. Section 2 describes the proposed method of vessel detection. Section 3 reports the simulation results while conclusions and future works are highlighted in Sect. 4.

## 2 Proposed Method

The proposed method is shown diagrammatically in Fig. 1. It consists of two main modules: the upper one does edge enhancement while the lower one serves the vessel extraction. The steps for blood vessel extraction using the proposed method are as follows:

***Step 1: Green Channel Extraction***—This paper uses the green color plane for vessel extraction due to the maximum contrast between the blood vessels and the background for the green channel retinal image [7].

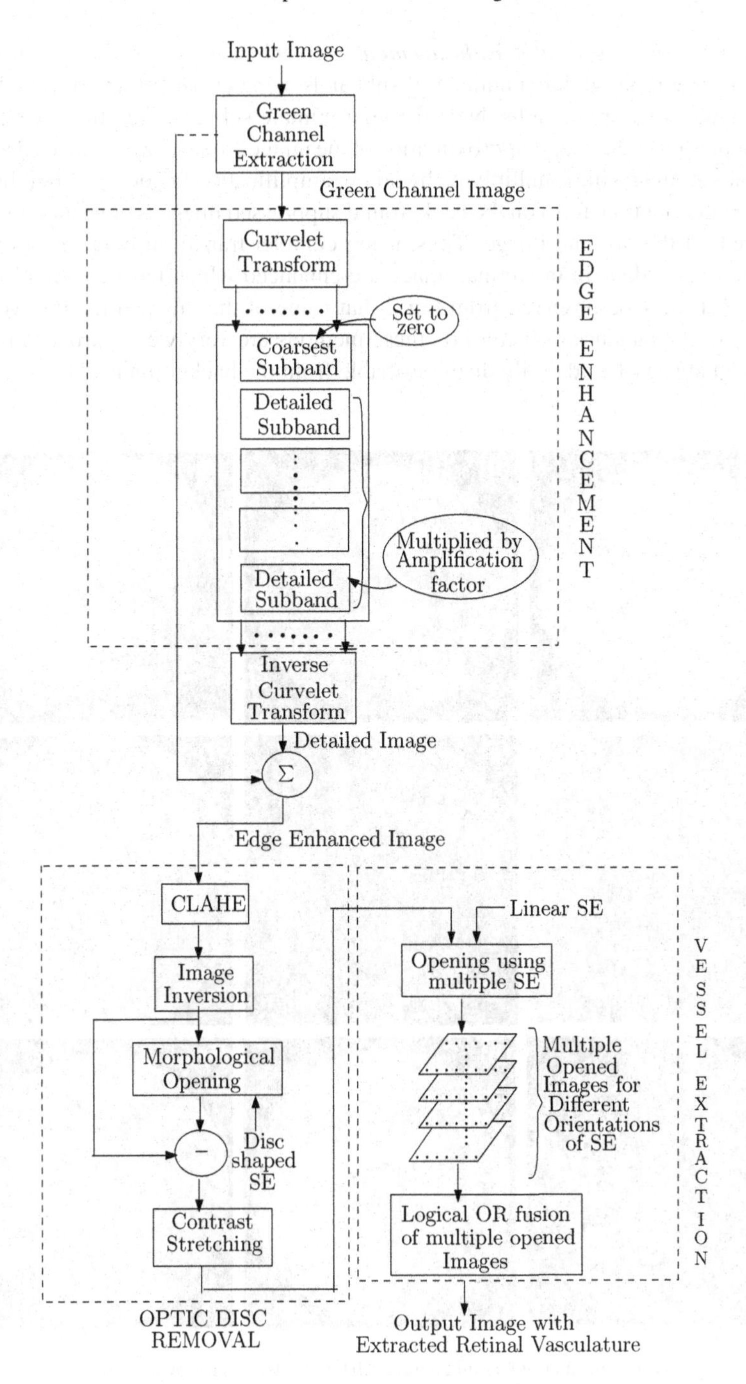


**Fig. 1** Schematic diagram of proposed method for retinal vessel detection

***Step 2: Curvelet based Edge Enhancement***—To enhance the vessel edges, first the image is decomposed into a number of subbands using curvelet transform with different scales and orientations. Next the approximate subband, i.e., the coefficients corresponding to the coarse approximation of the image are set to zero and the detailed subbands are intensified multiplying by proper amplification factor '$\alpha$'. After that the inverse curvelet transform on the background suppressed image is done and is superimposed on the original image. Thus, using curvelet transform both the strongest and the finest edges in the retinal image are enhanced which has been highlighted in Fig. 2. It may be observed from Fig. 2 that some of the tiny and the thin vessels which are not enhanced (visible) by these methods are very well sharpened by the proposed approach and easily distinguishable from the background.

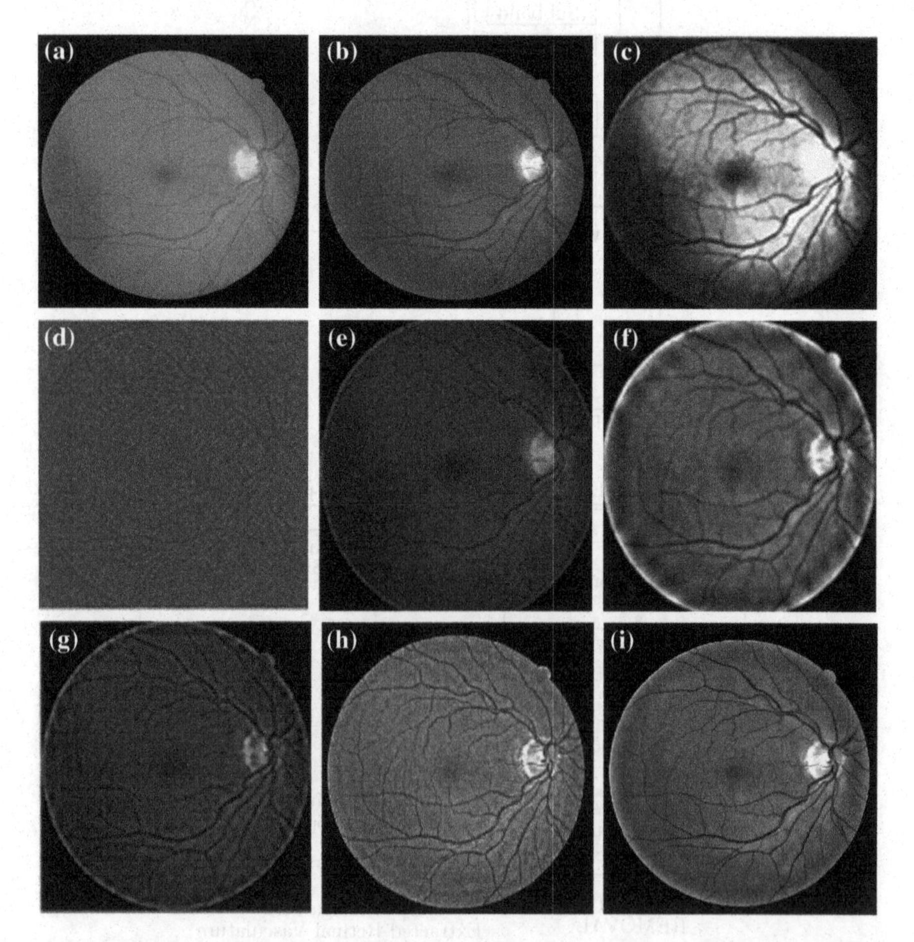


**Fig. 2** Simulation results. **a** Original retinal image. **b** Green channel image. Results of (**c**) histogram equalization, **d** local normalization, **e** unsharp masking, **f** wavelet, **g** contourlet, and **h** curvelet used in [7], respectively. **i** Curvelet-based edge enhanced image by proposed method

***Step 3: OD Removal***—Since the vessels emerge from the OD, OD removal is essential prior to vessel segmentation to extract the vessels that lie in the OD region. To accomplish this, the curvelet-based edge enhanced image is improved by contrast limited adaptive histogram equalization (CLAHE) and the corresponding inverted image is computed which is then opened using a disc-shaped structuring element (SE) with proper radius. This highlights the OD, eliminates the objects smaller in size than the SE and also smooths the background. The opened image is then subtracted from the inverted CLAHE image. In addition to OD elimination, the objects which were removed by morphological opening may also be retained back by this subtraction operation. To increase the poor dynamic range of the subtracted image, the lower intensity values between 0 and 0.2 are transformed into higher intensity values between 0 and 1 through contrast stretching by the 'gamma' ($\gamma$) function with $\gamma = 1$.

***Step 4: Vessel Extraction using multi-structure Morphology***—The two edges of a vessel are assumed to be parallel to each other and may be approximated as piecewise linear segments of finite width [2]. Therefore, opening the output image using linear SE of appropriate length will eradicate the objects whose length is smaller than the length of the SE. This is implemented in this paper by rotating the linear SE between 0°–180° at steps of 15° increment that produces a set of 12 opened images. Now these 12 images are combined by logical OR operation to get the complete vascular tree.

# 3 Simulation Results

This section presents the performance of the proposed method tested on all the images of the publicly available DRIVE database [12]. The values of '$\alpha$', the radius of the disc SE and the length of the SE are 2, 6, and 16, respectively and show good performance for identifying the different types of vessels obtained through multiple simulations over the DRIVE database.

The original images from the DRIVE database, the curvelet based enhanced image and the vessel structure extracted by the proposed method are shown in the 1st, 2nd, and 3rd row of Fig. 3, respectively. The extracted vessel map by different state-of-the-art methods and the hand labeled ground truth are presented in the 5th and 6th row of Fig. 3, respectively. Comparing the results of the 3rd and 4th rows with the manually segmented images, it is observed that the existing methods fail to extract many thin and weak vessels from the original image while the current method extracts the diverse types of vessel structures efficiently.

Performance of the proposed method and [7] which also uses curvelet transform for contrast enhancement followed by multi-structure morphology are compared for in Fig. 4 The thin vessels which are enhanced and are detected in the final segmentation result by the proposed method are marked by red boxes in Fig. 4c, e, respectively. This demonstrates the results that whereas the proposed method outperforms to find the diverse types of vessels, [7] fails to extract a number of weak and thin vessels.

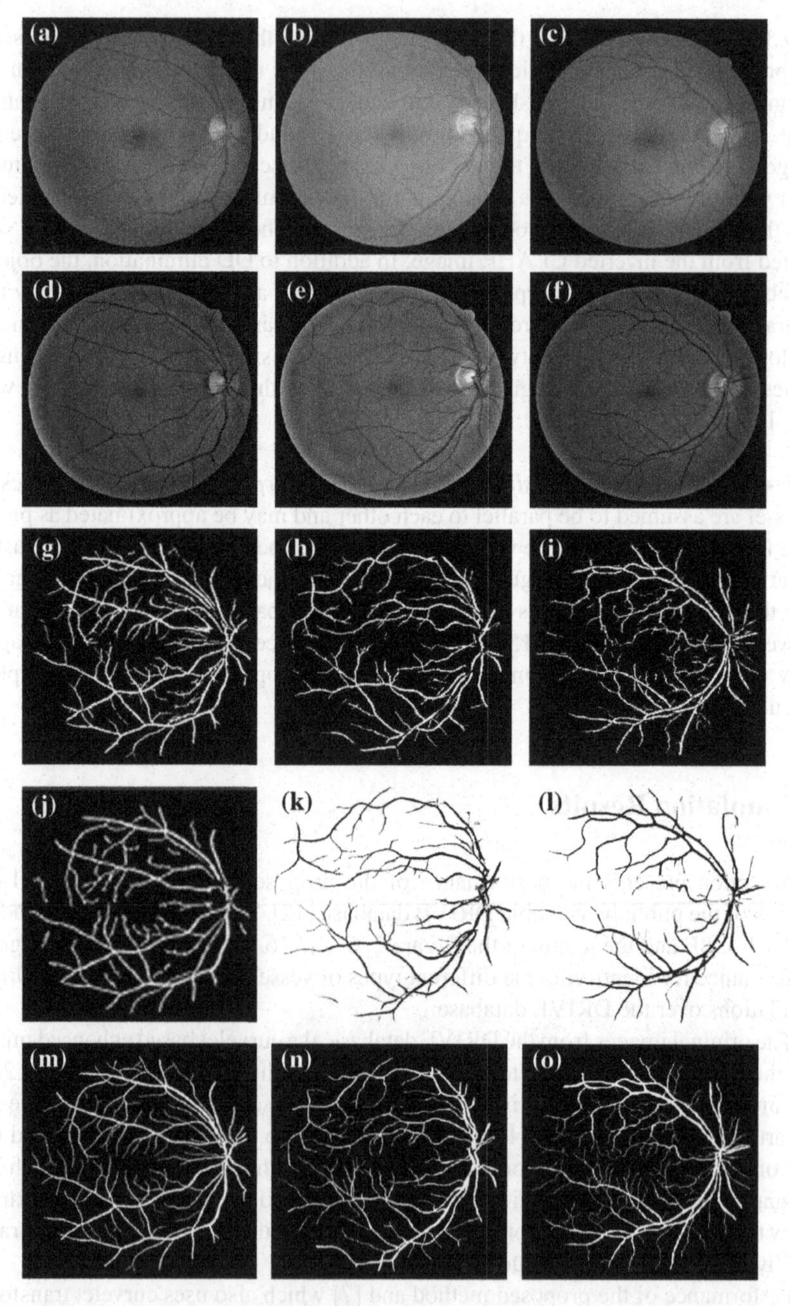


**Fig. 3** Simulation results. **a–c** Original retinal image. **d–f** Curvelet-based edge enhanced image. **g–i** Extracted vessel map by the proposed method. Vessel extraction results of (**j**) [11] and (**k**), (**l**) [9], respectively. **m–o** Manually segmented images

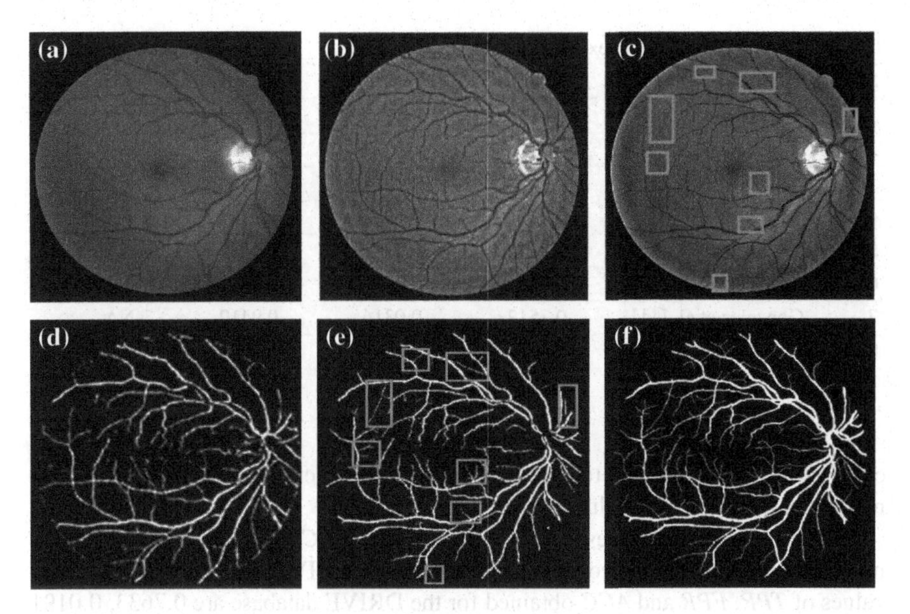


**Fig. 4** Simulation Results. **a** Green channel image. **b** Curvelet-based image enhancement by Miri et al. [7]. **c** Enhanced image by proposed method. **d** Vessel extraction results of [7]. **e** Extracted vessel map by the proposed method. **f** Manually segmented image

The metrices used to measure the performance of the proposed algorithm include (1) Detection Accuracy (ACC), (2) True Positive Rate (TPR) and (3) False Positive Rate (FPR). The Receiver Operator Characteristics (ROC) curve is shown in Fig. 5. The area under the curve (AUC) is 0.9706. This ensures high detection accuracy of the proposed method in comparison with the other methods. The slope of the ROC

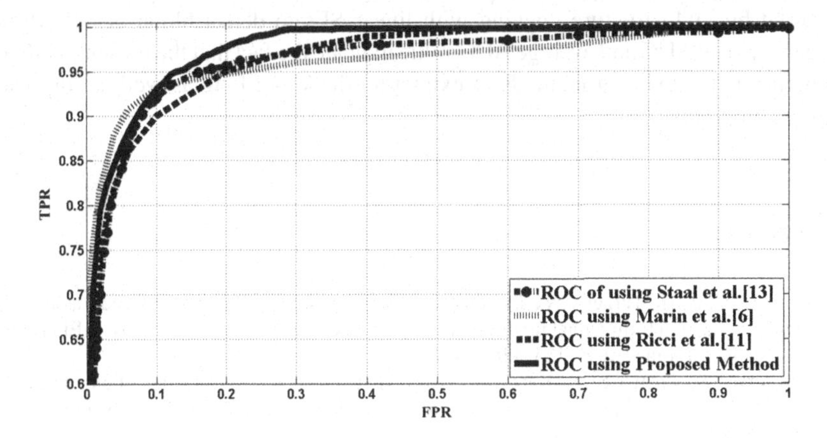


**Fig. 5** Comparative study of the ROC curves

**Table 1** Comparison of vessel extraction results

| Sl. No | Method | TPR | FPR | Average ACC | AUC |
|---|---|---|---|---|---|
| 1 | Chaudhuri et al. [2] | 0.6168 | 0.0259 | 0.9284 | NA |
| 2 | Ricci and Perfetti [10] | NA | NA | 0.9595 | 0.9633 |
| 3 | Niemeijer et al. [8] | 0.6898 | 0.0304 | 0.9417 | NA |
| 4 | Staal [12] | 0.7194 | 0.0227 | 0.9442 | 0.9520 |
| **5** | **Miri et al.** [7] | **0.7352** | **0.0205** | **0.9458** | **NA** |
| 6 | Marin et al. [6] | 0.7067 | 0.0199 | 0.9452 | 0.9588 |
| 7 | Gonzalez et al. [11] | 0.7512 | 0.0316 | 0.9412 | NA |
| **8** | **Proposed Method** | **0.7633** | **0.0191** | **0.9654** | **0.9706** |

curve is sharp as compared to the existing methods which indicates that the present method yields high TPR values for low values of FPR.

Table 1 presents the values of *TPR*, *FPR*, average *ACC* and *AUC* for the proposed method and the existing works on the test set of DRIVE database. The average values of *TPR*, *FPR* and *ACC* obtained for the DRIVE database are 0.7633, 0.0191 and 0.9654, respectively. Numerical results show that the average values of *TPR* and *ACC* are the highest and the value of *FPR* is the lowest for the present method

## 4 Conclusions and Future Directions

This paper proposes OD removal and blood vessel extraction for retinal images using curvelet transform with morphological image processing. Curvelet transform extracts the blood vessels of different structures and orientations with clear boundaries and curvatures. The OD is removed by morphological opening using disc-shaped SE followed by multistructure opening with linear SEs to detect blood vessel edges efficiently. The AUC and average ACC value indicate the fact that the present method outperforms the existing methods in extracting thick and long as well as tiny and thin vascular nets.

## References

1. Can, A., Shen, H., Turner, J.N., Tanenbaum, H.L., Roysam, B.: Rapid automated tracing and feature extraction from retinal fundus images using direct exploratory algorithms. IEEE Trans. Inf. Technol. Biomed. **3**(2), 125–138 (1999)
2. Chaudhuri, S., Chatterjee, S., Katz, N., Nelson, M., Goldbaum, M.: Detection of blood vessels in retinal images using two-dimensional matched filters. IEEE Trans. Med. Imaging **8**(3), 263–269 (1989)

3. Fraz, M.M., Barman, S., Remagnino, P., Hoppe, A., Basit, A., Uyyanonvara, B., Rudnicka, A.R., Owen, C.G.: An approach to localize the retinal blood vessels using bit planes and centerline detection. Comput. Methods Progr. Biomed. **108**(2), 600–616 (2012)
4. Fraz, M.M., Remagnino, P., Hoppe, A., Uyyanonvara, B., Rudnicka, A.R., Owen, C.G., Barman, S.A.: Blood vessel segmentation methodologies in retinal images—a survey. Comput. Methods Progr. Biomed. **108**(1), 407–433 (2012). http://dx.doi.org/10.1016/j.cmpb.2012.03.009
5. Goatman, K.A., Fleming, A.D., Philip, S., Williams, G.J., Olson, J.A., Sharp, P.F.: Detection of new vessels on the optic disc using retinal photographs. IEEE Trans. Med. Imaging **30**(4), 972–979 (2011). http://dblp.uni-trier.de/db/journals/tmi/tmi30.html#GoatmanFPWOS11
6. Marin, D., Aquino, A., Gegndez-Arias, M.E., Bravo, J.M.: A new supervised method for blood vessel segmentation in retinal images by using gray-level and moment invariants-based features. IEEE Trans. Med. Imaging **30**(1), 146–158 (2011). http://dblp.uni-trier.de/db/journals/tmi/tmi30.html#MarinAGB11
7. Miri, M.S., Mahloojifar, A.: Retinal Image analysis using curvelet transform and multistructure elements morphology by reconstruction. IEEE Trans. Biomed. Eng **58**(5), 1183–1192 (2011). May
8. Niemeijer, M., Staal, J., van Ginneken, B., Loog, M., Abramoff, M.: Comparative study of retinal vessel segmentation methods on a new publicly available database. Fitzpatrick, J.M., Sonka, M. (eds.) SPIE Medical Imaging, pp. 648–656. SPIE (2004)
9. Ramlugun, G.S., Nagarajan, V.K., Chakraborty, C.: Small retinal vessels extraction towards proliferative diabetic retinopathy screening. Expert Syst. Appl. **39**(1), 1141–1146 (2012). http://dblp.uni-trier.de/db/journals/eswa/eswa39.html#RamlugunNC12
10. Ricci, E., Perfetti, R.: Retinal blood vessel segmentation using line operators and support vector classification. IEEE Trans. Med. Imaging **26**(10), 1357–1365 (2007)
11. Salazar-Gonzalez, A., Kaba, D., Li, Y., Liu, X.: Segmentation of blood vessels and optic disc in retinal images. IEEE J. Biomed. Health Inform. **18**(6), 1874–1886 (2014)
12. Staal, J., Abrmoff, M.D., Niemeijer, M., Viergever, M.A., van Ginneken, B.: Ridge-based vessel segmentation in color images of the retina. IEEE Trans. Med. Imaging **23**(4), 501–509 (2004). http://dblp.uni-trier.de/db/journals/tmi/tmi23.html#StaalANVG04

# Author Index






R. Singh et al. (eds.), *Machine Intelligence and Signal Processing*,
Advances in Intelligent Systems and Computing 390,
DOI 10.1007/978-81-322-2625-3

Printed in the United States
By Bookmasters